Bad Decisions Make Great Stories

By
Pat Moore

Copyright © 2022

All Rights Reserved

Table of Contents

Dedication i
Acknowledgment ii
About the Author iii
Preface iv
A Huey in Peril 1
My Introduction to Skydiving 4
Running boards 27
Caving 28
Water Skiing 29
My Brief Pugilistic Career 33
Flying 36
Gordo 37
Never got shot from a cannon 38
Air Force 39
Exploding Thanksgiving 53
University of South Florida 54
Remedial Dating 63
Education 70
Jobs 72
Cars 74
Bodily Harm 76
Sports 79
Injuries 83
The Wolverine Bar 88
Vehicular mishaps 91
The Blues Brothers 93
Horse Race Handicapping 95
Stratonauts 96
Our Travel Channel 97
South Africa 104
Dubai 105
Airbags 107
Drone 109

Indoor Skydiving	112
Sailing	113
Our Hot Air Balloon Misadventure	115
Retirement (sort of)	117
Epilogue	119
Not quite the end	121
Never leave a six-year-old in the car unattended	122
Dad's mellowness is tested again	125
Sleep disorders	127
Bond, James Bond….	129

Dedication

My very tolerant and highly supportive wife, Penny

My extraordinary daughters Heather and Cait and their really cool husbands

And to our wonderful grandchildren, who should probably wait several years before reading this....

Acknowledgment

A word of thanks to a number of folks. Jack Carmody and his brother TJ encouraged me to commit these memories to paper.

Kudos to Alan Stark, Mike Marcon, Neal Fisher and Marc Cirigliano for their very helpful suggestions.

And a special mention to the individuals who have helped me enjoy so many activities and pushed me to do better: Skydivers Ski and Donna Chmielewski, Snowboard Racer Arden Sonnenberg, Ski Racers Rob Dexter and Ron Kapraszewski, Ski Racer and Golfer Jim Remy, all the members of Mountain Laurel Skiers Club and of course my partner in travel, photography, life and love, my wife Penny Trick.

About the Author

Pat Moore has been a TV weatherman, gymnast, springboard diver, ski racer, snowboard racer, college instructor, USAF airborne weather observer, computer geek, insurance salesman, small business owner, law student, rock climber, golfer, recruiter, web designer, photographer, videographer, artist, flyboarder, bungee jumper, unicyclist, barefoot waterskier, parachute rigger, pilot, skydiver and author. In 2008 he was the first person to win age group National Championships in both ski racing and snowboard racing. He's still not sure what he wants to be when he grows up.

His motto: "If you're not living on the edge, you're taking up too much room".

Preface

Everyone has adventures and misadventures. They can range from a mild speed bump to life-threatening situations. I've had more than my share of the latter in the past three-quarters of a century. I've experienced five parachute malfunctions, nearly gotten decapitated, fallen from a fast-moving truck, escaped from an apartment fire, almost got hit by a train, and was trapped in a cave. We're told that cats have nine lives. By my count, I've used that many.

What follows is a recap of some of the more foolish adventures. Sit back with your favorite potable and enjoy the ride…

* * *

A Huey in Peril

The iconic UH-1 Iroquois "Huey" Helicopter was the Army's workhorse for 42 years. One day in 1967, I almost managed to destroy one.

As a skydiving jumpmaster, I dropped hundreds of students without incident, but that safety record abruptly stopped, and the US Army almost lost a rather expensive chopper.

On this particular morning, the military club I belonged to arranged for sport jumps in an empty field near Sun City, Florida. I was jumpmastering six or seven students, and I planned to drop one per pass over the field. In the waist-gunner spot on the left rear of the chopper was a civilian friend named Bob Branch, whom we had smuggled aboard. Instead of a co-pilot, that seat was occupied by a nurse from MacDill AFB. The pilot on a Huey sits in the right seat. As I recall, his name was Travis, and he had flown many jump runs for us in the past.

The jump students all sat on a wide bench seat. As we climbed to 3000 feet and neared the exit point, I would have the student in the far left seat stand up, and I'd hook up the static line that would open his parachute. The first student that day was named Enrique, who was making his second jump. Second jump students concerned me more than first-timers. If it's your first jump, you're usually too scared not to do exactly as told. When your second jump rolls around, you know what to expect. Squirrelly behavior is not uncommon.

I had him stand in the door with his left foot poised on the edge. I gave Captain Travis hand signals to guide us over the exit point, then I slapped Enrique on the leg and yelled, "GO"! He appeared to be trying to exit but wasn't going anywhere. He acted as if he was caught on something. I signaled the pilot to go around again and sat Enrique down. I asked, "What happened"? He replied, "I thought we were past the spot." I said, "That's my job! Get ready to go again". Big mistake. It turns out that he didn't go because he was terrified.

He assumed the exit position again, and I slapped him on the leg. He seemed to be fighting himself, and then his left foot slipped out into space. He tried to throw himself back into the helicopter but instead landed on his stomach on the landing skid. His chest-mounted reserve chute burst open, and the vortices created by our forward progress caused the chute to wrap around the skid several times before inflating. That would have been enough to bring down the helicopter, but most of the suspension lines snapped from the incredible forces placed on them. Unfortunately, the trailing lines were dangerously close to the tail rotor. I've drawn this pencil sketch to give you an idea of the situation.

The chopper was shaking violently, and we were rapidly losing altitude. Enrique was literally tied belly up from the skid and bleeding from the neck. Bob and I climbed down and couldn't free him. Captain Travis had his hands full trying to keep the aircraft under control and screamed, "Get that boy off my helicopter"! "I need a knife"! I yelled, and he bellowed, "My flight suit!"

I dove across the flight deck and found a switchblade in a pocket between his thighs. I opened the knife right between his legs, and I swear he levitated right out of his seat. By now, we were down to 800 feet, and we were close to making national news in a bad way. I handed Bob the knife as I climbed back down on the skid. Bob began sawing on the remaining lines. Enrique was conscious but in a lot of pain. We learned later that his left shoulder had been dislocated by the impact. As Bob cut the last line, Enrique fell away. All this time, he had thought it was his main parachute that had deployed, and, in his mind, he knew he would have to pull his reserve when he finally separated from the Huey.

I will never forget the look on his face as he realized he had no reserve. His static line successfully deployed his main chute, and he descended to the ground with one arm hanging limply. I then realized I was standing on one foot on the outside of a helicopter holding on with one hand at an altitude too low to jump, and scrambled back aboard. The pilot followed Enrique down and landed near him. We carried him on board and flew him to the hospital at MacDill AFB, where he made a full recovery.

Since we had very nearly killed more than a half dozen people and had come close to destroying a very expensive helicopter, the military insisted on a full accident report. The problem was that they wanted a statement from the serviceman in the waist-gunner slot who had been so helpful. Obviously I couldn't get a statement from Bob so I enlisted the help of another serviceman who wasn't even there that day. I said, "Ross, I'm going to write a report and you're going to be a hero. You just have to sign it". He had some trepidation but a bribe of a six pack won out and the incident was finally put to rest. As you might expect, Enrique never made another jump.

* * *

My Introduction to Skydiving

I turned 16 in October 1962, and Dad asked if I wanted to jump out of an airplane. "You bet!" I replied without hesitation. Dad was stationed at Westover AFB in Chicopee, MA, and Parachutes, Inc., one of the nation's first skydiving schools, was located nearby in the town of Orange. The sport was really in its fledgling days, and practitioners were mostly ex-Airborne types from WWII. December 2, 1962, was the last day the school was going to be open until Spring, so we headed north early that morning. Our small class of first-time jumpers sat on bench seats outside as we watched our instructor, Lew Sanborn, descend under an orange and white surplus military chute that had been modified to make it somewhat steerable. He wore a canary yellow jumpsuit and looked really cool! There were a couple of NYC-based secretaries in our class, and I overheard their comments on this sky god who had just landed in front of us. I decided then and there that I, too, would become proficient at this exciting sport. Author's note: I attended a US Parachute Association event in 2015 and reconnected with Lew. More than 40,000 D (Expert) skydiving licenses have been issued, and Lew Sanborn is D-1. Now in his nineties, the man was still jumping as late as 2020!

I won't bore you with the details of our training that day, but when it came time to step out of the plane, I was terrified. Granted, a static line attached to the plane would open my chute, but that did little to assuage the fears of a kid barely old enough to drive. All the practice exits I had performed so flawlessly were for naught as I tumbled out of the plane in a fetal position. The canopy deployed perfectly, and I landed without incident in a huge target area in the middle of the airport. I was hooked!

We moved to Suitland, MD, that summer, and there was no opportunity to jump. My senior year of high school was beginning, and I was considering college. In September 1964, I enrolled at the University of Florida. Armed with a 1937 Remington Rand portable typewriter, I arrived in Gainesville ready to take on the world. This was so long ago that the campus cops drove Studebaker Larks.

Shortly after school began, I spotted a flyer about a skydiving club and immediately contacted the organizers. Like me, the club's president, Dave Henson, was very new to the sport.

Florida has a large number of former WWII bomber training bases, and the nearest to Gainesville was in the town of Williston. In February 1965, we headed there and met a guy named Ron, who was running the school. He didn't exude the same confidence as Lew Sanborn, and his operation was far less sophisticated than I had encountered at Orange. We were introduced to the pilot named Homer Satterfield. Homer looked as though he might have flown in the First World War. As we discovered later, he knew nothing about jumping but was perfectly happy to take us up and drop us wherever we wanted.

We did a few practice parachute landing falls and observed Ron packing our chutes. He took us up, dropped a wind streamer, and had us exit. The thrill was as exciting as I had remembered two years earlier. At least this time, I managed a decent exit.

Dave and I made a few more jumps and graduated to free falls. No longer tethered to the plane, we exited and pulled our ripcords. We were pretty full of ourselves. We had arrived. We were skydivers!

We headed back to the airport a week later, but Ron wasn't around. Homer said he'd be happy to take us up. The chutes weren't packed, but we were pretty sure we remembered how Ron had packed them a week before. Dave's lines were so tangled that he had to disconnect the Capewells, the pieces of hardware that connected the canopy's risers to the harness. He got everything reassembled, and we got the containers closed. We then headed to the Cessna 172, where Homer had removed the passenger door. While climbing, I rehearsed what I needed to do. *Climb out on the step, leap off with my body in a big arch, count to three, and pull the ripcord.* Homer asked us where we wanted to exit as we reached jump altitude. It occurred to Dave and me that we had no clue how to pick the exit point. We hadn't brought along a wind streamer to drop, and I doubt we could have figured out to use it. I pointed to a spot over the end of the runway and said, "I think that's where we got out last week."

Of course, that would only be helpful if the winds aloft were exactly the same, and, as it happened, they weren't. I climbed out anyway as Homer put his foot on the brake pedal so I could stand on the right tire while hanging on the wing strut. I recall shouting "One thousand, two thousand, three thousand," then realizing I was still clutching the plane. I let go and began the count again. A quick yank on the ripcord opened the backpack, and a small spring-loaded pilot chute pulled the main canopy out. It fully inflated, and I was elated. My pack job worked!

I looked up to see Dave, and his canopy was open too, but something didn't look right. When packing the parachute, he had detached the risers and had a 50/50 chance of reconnecting them correctly. He guessed wrong. The parachute was backwards! Instead of moving slowly forward, he was heading in a direction he couldn't see. He also discovered that pulling down the steering toggle had opposite the desired effect – pulling the right one turned him to the left! Dave was an Engineering major and a pretty bright guy, and he was able to adapt. He managed to steer to an open spot on the field and make a reasonably tidy landing. I, on the other hand, had been so mesmerized by Dave's travails that I paid almost no attention to where I was going. I managed to miss the entire airport and came down in a lone tall tree. Fortunately, I wasn't dangling too high off the ground and could get out of the harness without any difficulty. I scampered up the tree and retrieved the canopy, only to discover a number of tears produced by tree branches.

Dave and I walked back to the packing area and examined the canopy. We found some duct tape and patched the tears. We then reattached Dave's risers properly and repacked the canopies. Homer took us back up, and this time everything went smoothly. We were now experienced pros!

Dave bought all the equipment from Ron and started giving lessons. We moved the operation to Kay Larkin field near Palatka, FL, and I experienced my first malfunction on my 11th jump. Upon opening, a suspension line had flipped over the top of the canopy, creating an aptly named "Mae West." I deployed my reserve successfully and wasn't dissuaded from sticking with the sport. Exactly fifty jumps later, it happened again, and I made a standup landing on a World War II surplus twill emergency chute. In later years I had two total malfunctions – the main didn't open at all! Over a fourteen-year skydiving career encompassing more than 2000 jumps, I had to ride my reserve down five times.

***

In the summer of 1966, Dad decided he'd like to try skydiving. He headed out to historic Manassas, Virginia, with my 18-year-old sister Jan and my 16-year-old brother Mike. They all signed up for lessons. Our little sister Kathy was only six, and she was content to hang out and watch. Manassas was home to two sport parachuting clubs – the Targeteers and the Pelicans. The

Pelicans were as rowdy a bunch as you'll ever run into. They jumped hard and partied harder. One member told me that he had ridden down a double-entanglement and slammed into the ground behind a shed. As he lay there with a few broken limbs, he heard his club members running around the shed to rescue him – or so he thought. As soon as they got in earshot, he heard, "I got his boots"! I got his helmet"!

One of the Pelicans entered the outhouse near the target, and I watched as another member sneaked up to the door. He pulled the pin on a red smoke grenade and threw it inside. Incredibly the guy inside didn't come out immediately. He went about doing his business and emerged in a huge cloud of billowing red smoke. Instead of getting mad, he just said, "Excuse me, folks. Sure glad I didn't do that in the house!"

The Pelicans' pilots were every bit as whacked as the jumpers. On more than one occasion, I watched a Cessna 182 descend after discharging the jumpers and level off at ground level, flying toward the landing target. There was a line of telephone poles at the end of the drop zone, and the pilot would fly between the poles and *under* the phone lines with little clearance. In Ridgley, Maryland, a few years later, I saw the stunt repeated, but this time the pilot had to clear a fence that was under the lines. Amazingly the planes and the pilots always emerged unscathed.

The entire Moore family progressed to freefalls and were well into the sport when the drop zone experienced three fatalities in rapid succession – the last one was an 11-year-old boy. The rest of the family retired from skydiving, although Mike did make a few jumps years later.

After Air Force Tech School, I was assigned to my first and only duty assignment at MacDill AFB in Tampa, Florida. The Tampa Skydivers had a very active club in the nearby town of Zephyrhills, and I couldn't wait to become a member.

Over the next several years, I accumulated more than 2000 jumps, some of which stuck out in my mind. We did a lot of air shows and demo jumps back then (mainly because free jumps enticed us), and sometimes the landing areas were tighter than we would have liked. My accuracy had always been pretty good, but I created my own problems. With a background in gymnastics, I liked to show off under the canopy by doing an inverted "Iron Cross", hooking my feet in the lines over my head, and drifting down headfirst. I would always slip my feet back down just in time to make a crowd-pleasing one-foot standup landing. At one air show in Lakeland, FL, I was doing just that when I had a small problem. All of us jumpers were wearing brackets on one foot holding smoke bombs that enabled the spectators on the ground to see us.

When I slid my feet through the lines over my head, the smoke bracket entangled with the lines! When I realized what I had done, I struggled to extricate my foot to no avail. Realizing I was about to put my Bell helmet to the test, I forced myself to calm down and slowly worked my foot out the snarl. My legs swung down just in time, and the crowd went crazy, thinking it had all been intentional.

Armed Forces Day was always a fun time to make exhibition jumps. The crowds were huge and treated us like sky gods. I had sprained my ankle badly on a jump a week before one of these events and was grounded for a few weeks, but I convinced the organizers that I could land well away from the crowds. The helicopter pilot was a friend, and when I did a one-foot standup landing at the end of the runway, he landed nearby to pick me up. Inside the chopper were my crutches—anything for a free jump.

Normally on Armed Forces Day, I would do the Intentional Cutaway. Five of us would exit the chopper at 7500 feet with smoke bombs streaming from brackets on our ankles. The other four would freefall down to 2500 feet and open their canopies. I would open shortly after leaving the chopper so I could float more than a mile up. The announcer was supposed to tell the crowd that I was wearing three parachutes instead of two and would cut away from one and go back into freefall.

The PA system failed, and the crowd was stunned to see me slip away from my inflated canopy and begin tumbling. One guy in the stands screamed out, "He done fell outta his parachute"! I'm told people were screaming and clutching their hearts. The other four guys had opened a little high, and from the angle the crowd was looking at, it appeared as though I was plummeting right through one of the open chutes. I finally opened it, and the PA system came back on with the announcer saying, "Folks, that was *supposed* to happen"!

While competing at a skydiving meet in Indiantown, FL, I had the bright idea of having the pilot fly low and slow over the St. Lucie Canal. I'd jump into the water and have a boat pick me up. That way, I could say I had jumped from a plane without a parachute. Before that could take place, a few jumpers jumped into the canal with parachutes, and one of them drowned. That put an end to my attempt.

At Zephyrhills, the Tampa Skydivers offered lessons to first-time jumpers. I gave a few lessons, but mostly I acted as a jumpmaster. That involved doing a preflight gear check. As I examined the harness of the male students, I'd invariably say, "Turn your head and cough." If they complied, I knew I had a student who would do exactly what I wanted when we got to altitude. Blind obedience is a beautiful thing. If they just laughed, I figured I'd need to keep an eye on them. Obviously, this didn't work on the female students. Speaking of female students, I had one actually fall asleep during the climb to jump altitude. That unnerved me. Once in a while, I'd have one who refused to jump. That was fine. They'd just ride the plane back down, and I'd assure them there was no shame in it.

On more than one occasion, a large group of hungover guys would show up on a Saturday morning looking for lessons. I'd address them with something like this: "I'm guessing you all were at a party last night, and someone brought up the idea of jumping out of an airplane. You all agreed to do it. I'm also guessing some of you are having second thoughts. This sport isn't for everyone, and if your enthusiasm is waning, now is the time to take a pass. No one will think less of you for it. And I'd prefer you to change your mind now instead of when we're at 3000 feet." Invariably one or two would take a pass.

Occasionally a student would be outside the plane and then refuse to jump. Getting them back inside was a little tricky, but it could be done. On one occasion, though, a woman froze and wouldn't climb back in. In later years we would mount a step over the plane's right wheel, but in the early days, we simply had the pilot hold down the brake pedal to keep the wheel from turning. On this particular day, we were heading upwind at 3000 feet, and the student was sitting on the plane's floor with her back to the instrument panel (the passenger seat, control yoke, and the right door had been removed). Her static line was securely anchored to the pilot's seat. I was on my knees with my head out the door giving the pilot thumb signals directing him to where I wanted the student to exit. I yelled, "Cut," and he throttled back to cut down on the wind speed she'd encounter climbing out of the plane.

With reduced power, the plane can't maintain altitude and will eventually stall, so you don't want the student taking too much time. I yelled, "Sit in the door!" and she obeyed. She put her left hand on the wing strut, placed her left foot on the tire, and then swung out, grabbing the strut with her right hand. She had her eyes on the horizon and was in perfect position. I yelled "GO!" and she yelled "NO!" I repeated the command, and she repeated the response. I said, "Okay, get back in," and she again shouted, "NO!" I had experienced pretty much everything a student could do, but this was new for me.

The pilot was on the verge of a stall, and we were well past our exit point. I suddenly thought of a solution. I hollered at the pilot, "Take your foot off the brake!" He nodded and grinned. The wheel began to turn, and the girl got a shocked look on her face. She managed to hop her other foot on the wheel as the first one slipped back and started running like a hamster in a cage. It didn't take long for her to totally slip off and, with a shriek audible to those on the ground, dropped away from the plane. Her parachute immediately opened, and she drifted down. There was no way she would land at the airport, but she managed to touch ground unscathed on an adjacent golf course. She gathered up her canopy and strolled back to the drop zone. I dropped my other two students and then jumped myself. Later on the ground, I walked up to her with trepidation, but she had a big smile and hugged me. She thanked me, and I was relieved when she decided not to try a second jump.

* * *

Remember those old black and white TV commercials with John Cameron Swayze? He'd affix a Timex watch to a boat's propeller, run it at full speed, and then show that the watch still worked. "Timex. Takes a Licking and Keeps on Ticking" was the slogan. I should have submitted my watch for a commercial. I had been jumping at Zephyrhills all day. I then retired for the evening to the Tampa Skydivers clubhouse just off the airport to do some canopy repairs on the sewing machine (by then, I had decided that duct tape probably shouldn't be used). I looked down at my wrist to see what time it was and noticed that my watch was missing. My dad had given me a Timex Electric upon graduation from high school, and I sure didn't want to lose it. As near as I could recall, I had been wearing it on the day's last jump, and the exit point was over the adjacent golf course.

The next morning I walked to the pro shop and inquired if anyone had found a watch. The Golf Pro asked, "What time did you lose it"? I guessed it was a little after 2 p.m. He said, "How about 2:17? You almost killed one of our golfers"! He held up my watch and, sure enough, it was stopped at 2:17. I shook it a couple of times, and believe it or not, it started ticking again! I didn't want to risk losing it again, so I designed a watchband made of multicolored parachute webbing and Velcro. They became so popular that I kept a sewing machine in my van and made custom watchbands right at the drop zone.

* * *

I enjoyed "relative work," where many jumpers formed formations in freefall but preferred the individual events of style and accuracy. One day I grabbed a ride on a DC3 that was taking a load of relative workers to 12,500. I asked the pilot to make a pass over the field at 6,600 feet so that I could get in a practice-style jump. He agreed but said he wouldn't throttle back for my exit. As we approached the exit point, I was clowning around and did a high back flip out the door. It wasn't until the rest of the jumpers joined me on the ground that I learned that I had barely escaped decapitation. Apparently, my exuberant upward jump combined with the higher than normal forward speed resulted in my head barely clearing the plane's horizontal stabilizer. Some things you're better off not knowing.

* * *

The United States Parachute Association held the National Championships in different areas around the country. I competed at Marana, AZ, Plattsburgh, NY, and Tahlequah, OK. It was at the last spot that I learned how painful rib injuries could be. The event ran over several days, and competitors competed in accuracy and style events. In accuracy, the target was a disk 10 centimeters across (less than 4 inches), and you were measured on your first point of contact. The style event was a little more complex. You left the plane at 6600' on a downwind run and aimed your head straight down. I'm here to tell you that Galileo was wrong. His calculation of 176 feet per second goes out the window when a human being is flying head first toward the ground. We could go much faster and would do so in order to build up speed going into the first of a series of six turns and loops that the judges would grade us on.

I was trying very hard to make the US Team at this particular competition. On an accuracy jump, I miscalculated and was going to miss the disc. I wrenched my body as hard as possible and jabbed my foot down a microsecond before my shoulder blade slammed into the pea gravel covering the target area.

Pea gravel is a terrific substance for sliding in with your feet but not as forgiving on body slams. I had the breath knocked out of me but didn't think I had done any permanent damage. The next day I was proved wrong.

Early the following morning, I suited up for the style event and clambered aboard a Britten Norman Islander twin prop plane to begin the climb to jump altitude. The pilot was dropping one style competitor per pass and was making some tight rotations. I was on my knees on the floor of the plane, and I recall the G forces were pretty intense. As I tightened down the belly band of my reserve parachute, I felt a twinge on my chest. I must have had a strange look on my face because the next jumper asked if I was all right. "I think so," I replied. We turned on jump run, and my number was called. I exited, dropped my hands to my side, and pointed my head down. Fifteen seconds into freefall, I slammed my knees into my chest and stuck out my left hand, trying to initiate a quick left-hand 360-degree turn. I followed that with a quick right turn and then a back loop. It was in the middle of the back loop that a portion of my rib separated from the cartilage. The pain was incredible. I felt as though someone had taken a sharp poker out of a fire and stabbed it into my chest.

Skydivers will recognize the acronym NSTIWTIWGD. It translates to "No sh*t. There I was. Thought I was gonna die". My body, still falling at over 120 mph, went limp, and I fell out of control. The judges who were looking through powerful binoculars called telemeters realized I was in trouble. I needed to pull my ripcord, but I knew the pain of the opening shock would be excruciating. Recognizing that pain was still preferable to impact with the ground, I gave a yank of the ripcord. My parachute opened, and I almost blacked out from the pain of the opening shock. The pressure being placed on my chest by my reserve parachute was more than I could stand. With the main canopy successfully deployed, I no longer needed the reserve, so I unhooked its right attachment point and let it hang down to my left. I couldn't raise my right arm, so I steered by pulling the toggles with my left hand only.

In the days before square canopies became popular, my round parachute (a French model called a Papillon) was backing up in a strong wind. I impacted the ground and fell back. The slung reserve hit the ground and bounced back, crashing on my injured rib. The next thing I remember was being loaded on a stretcher and put in the back of an ambulance. It was one of those old models that were indistinguishable from hearses. My wife Connie and a few friends hopped in with me, and we headed for town. I thought it was safe to assume that the ambulance driver had an inkling of the location of the hospital, but that assumption was dashed when I heard him ask a roadside construction crew for directions. One of the workers asked, "Which hospital? The White one or the Indian One?" I swear everybody turned to look at me before answering. I later discovered that the driver was the pilot's sixteen-year-old son, who had just gotten his driver's license and had volunteered to drive the ambulance. To his credit, he managed to get me to the hospital.

The back door of the ambulance opened, and my friends slid out of the stretcher. It was one of those wheeled models that telescope up and locks in place once you get it out of the vehicle. They yanked it up and let go. Unfortunately, the locking mechanism only engaged at my foot end. The head end slammed back to the pavement, and I slid off the stretcher head first. I heard "Oops" and "Sorry about that" and then found myself in the X-ray room.

There's not much that can be done for rib injuries. Nowadays, they don't even wrap the ribs because it restricts deep breathing, and there is a danger of contracting pneumonia. Back then, I was fitted with a giant elastic band secured with Velcro and told to take it easy for six weeks. I withdrew from the competition and watched my wife finish just one spot out of the Women's US team.

When the competition ended, we set out in our 1970 Maverick across the deep south from Oklahoma to Florida. We had just crossed the Mississippi River and entered Tennessee when I offered to drive. I wasn't supposed to be driving, but it was all Interstate driving, and I figured Connie could operate the gear shift until we got up to highway speed. Things started smoothly, but we weren't on the road five minutes when the right rear tire had a blowout at 70 mph. I clamped both hands on the wheel and ignored my protesting ribs as I nursed the car and its three functional wheels to the breakdown lane.

We sat in the car for a moment and just looked at each other. Without a word, we opened the trunk and began hauling out cases of Coors beer to get to the spare tire. Coors wasn't available east of the Mississippi back then, and it was illegal to transport. I have no idea why this was so, but a truck-driving friend was arrested for bringing back a couple of six packs. We must have had more than a dozen cases in the trunk. We carefully laid them on the side of the road and stacked our parachutes and suitcases on top of them to hide them.

Ford's idea of a jack and lug wrench left a lot to be desired. Because the wheels were recessed so deep in the fenders, the lug wrench barely cleared the car's body. To make matters worse, it wasn't even at right angles! The angle was so obtuse that it was difficult to get any torque on the lugs. When your ribs are screaming at you, the job doesn't get any easier. Eventually, we hit on a plan where we'd put the wrench in a position roughly parallel with the pavement and stand on it. Bouncing up and down finally freed up the lugs. All the while, we were watching the road hoping Officer Friendly *wouldn't* show up to help. Forty-five minutes later, we managed to remount the spare and repack the trunk. I didn't offer to do any more driving. We eventually made our way back to Tampa, and those cases of Coors eased the recovery. To this day, I have a bump on the edge of my ribs as a reminder of that trip to Tahlequah, Oklahoma.

* * *

I mentioned that I had experienced two canopy malfunctions early in my jumping career and had to resort to the reserve or "back-up" chute. In my entire jumping career, I had to ride the reserve down five times, the latter three for very different reasons.

Jumpers occasionally dropped their ripcords after opening, so they took a lead fishing weight and crimped it on the ripcord between some of the pins that held the backpack closed. The weight prevented the ripcord from slipping completely through its housing. I installed mine above the top pin, and when I attempted to pull the ripcord, the lead weight jammed against the housing and wouldn't allow me to pull it far enough to free the pins. Imagine falling at 120 mph and tugging on a handle that just won't budge. At 2000 feet, you have 11 seconds to impact with terra firma, so you don't want to spend too much time fighting the problem.

I rolled over on my back and yanked the handle on my chest-mounted reserve parachute. The opening shock was sudden and brutal. Something hit me in the back of the head. As painful as it was, it sure beat the alternative. I rode down the reserve and discovered what had caused the pack to remain closed. I then realized my boots had hit the back of my helmet. My body had been bent double by the opening shock! I fixed the problem with the ripcord and a few months later had a problem of a different sort.

Connie and I competed in a team event called "relative work" with Ski and Donna Chmielewski. They are a remarkable couple who have logged thousands of jumps and still jump after all these years. For a while, they toured for Budweiser doing exhibition jumps as the "The Flying Eagles."

On this particular day, we had exited from 7500 feet and began creating a formation we had been practicing. We would fall at terminal velocity for 30 seconds performing the routine, and then open. Midway through, Donna broke ranks and maneuvered over my backpack. She immediately spun around and "flew" over my backpack again. I had no clue what she was doing, but we were running out of altitude, so I reached for my ripcord to pull and discovered it *wasn't there*! The stitching that held the housing to my harness had broken free, and the ripcord had fallen out of its pocket. Donna had seen it trailing above me in freefall and had tried to grab it and pull it for me. I made a vain attempt to reach behind my head to grab it and realized that wouldn't work. Gritting my teeth, I rolled over on my back and fired my reserve again. Happily, the reserve worked fine once again.

The fifth and last descent on a reserve was a bit bizarre and had never happened to anyone in the sport up to that time. We had progressed from round parachutes to square – the ram air "flying mattresses" you see today. The forward speed was remarkable and allowed us a much higher degree of accuracy when landing. Instead of landing downwind in the target area as we did with round parachutes, the square chutes could penetrate the wind, so we landed upwind. The chest-mounted reserve blocked my view of the target, so I came up with a solution. I designed a harness with an extra attachment point for the reserve low on the left side of the harness. Once the main was open, I didn't need the reserve, so I'd disconnect it and reconnect the reserve's right attachment point on my left hip. The left attachment point just hung out in space.

Never one to be content to just ride a canopy down, I discovered that pulling down on one of the front risers would cause the parachute to dive forward and begin spinning rapidly. The parachute would race downward so fast that I could actually see the horizon *above* the canopy. Recovering from this spiral was simple – just let go of the riser. My reserve chute was equipped with an automatic opener that was designed to deploy if I passed through 1000 feet altitude at speeds greater than 2/3 terminal velocity. Apparently, my earthward spins triggered the sensor, and the reserve (remember it was hanging on my left side) suddenly burst open, inflating the 26' canopy. I instantly realized what had happened and thought, "No problem, I'll just ride both chutes down."

That would have been fine under a round parachute, but no one had ever deployed a reserve with a square before. The forward speed of the main chute was so powerful that it continued going forward, fighting the drag of the reserve. It got level with me in front and then dove toward the ground collapsing and floating back up at me. Since it could entangle with my reserve, I immediately popped the canopy releases to jettison it. Just as I did, I realized that I was only connected to one side of my reserve! Fortunately, when I built my reserve, I had sewn a connecting strap between the reserve's two connectors that attached to the canopy. It was standard safety practice in case a reserve ever became detached on one side. To the best of my knowledge, no one had ever needed one before! The strap was a meter long, so I was only attached at my left hip to the right side of my reserve. The left side of the reserve was a full meter higher, secured by the strap. The reserve started a slow spin with me hanging below it, powerless to steer it. I crashed into a tree and slammed into the ground, twisting my ankle. Here's a rough sketch of my predicament before I jettisoned the main chute on the left. In retrospect, I guess I was lucky. The Darwin Awards weren't created in 1976, but I came close to being an early contender.

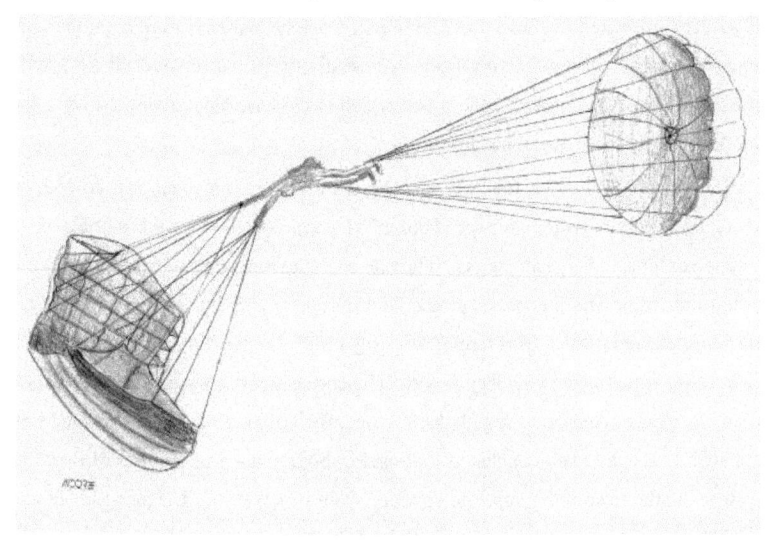

Whuffos are spectators. Their name comes from "Whuffo you jump from a perfectly good airplane?" In all my jumps, I never saw a "perfectly good airplane". If you had seen the relics we used to get to altitude, you'd agree. Jump planes take a lot of abuse, take off from some bumpy runways, and carry more occupants than they were designed for. The Cessna 182 is a real workhorse, but the one we used at Riverview, Florida, was pushed too far one day. Connie and I had just completed a milestone – her 1000th jump and my 2000th. It was the last jump of the day and went off without a hitch. The problem came the next morning. Mac McGraw took off with a load of three jumpers. Normally he'd carry four but was running light. As they approached jump run at 7500 feet, the jumpers crowded out on the right wheel with a small platform built over it. Without warning, the wheel strut snapped off at the fuselage, dropping the three jumpers unexpectedly. Falling with them were the wheel and strut. Astonished, they managed to avoid the falling debris and opened their chutes. They watched the wheel slam into an empty field below them.

Mac, meanwhile, was stunned to realize that his plane with tricycle landing gear now had "bicycle" gear. He was left with a nose wheel and a left wheel. How the heck was he supposed to land? Getting Mac worked up about anything took a lot, but this had his undivided attention. He chewed on his cigar, keyed his radio mike, and called the tower at Tampa International Airport. "Fellas, you're not going to believe this," he began. Once he had convinced them that, indeed, a vital component of his plane was conspicuously absent, they contacted MacDill AFB, which had the capacity to foam their runway. Mac was directed to the base's main runway. Once the foam had been put in place, Mac began his descent.

A terrific pilot, he delicately touched down on his left wheel only and coasted with the nose gear up as long as possible. When his speed dropped too much to maintain this attitude, he lowered the nose and continued his balancing act. When he could no longer keep the right wing up, he let it drop into the foam. The plane executed a slick pirouette and came to rest. An examination revealed minor damage, which Mac termed "runway rash". The FAA took a strong interest in this incident because it had never occurred before. The spring steel in landing gear is a continuous piece from one wheel to the next through the fuselage. By design, it's built to withstand some very hard landings. The examiners concluded that having bodies stand on the top of the wheel was placing stress on the strut in the opposite direction from which it was designed to handle. I don't recall ever hearing of a similar occurrence, and happily, no one was hurt.

* * *

Problems with jump planes weren't always the plane's fault. Deland, Florida, was the home of a large skydiving center that catered to jumpers who liked to build formations in the air. As many as 30 would exit a DC-3 at once for 60 seconds of freefall. Beginning life as the military C-47, the workhorse of WWII enjoyed the longest production run of any plane ever built and evolved into the DC-3 passenger plane. A twin prop tail dragger, it was ideally suited to hauling large groups of jumpers.

Deland could get very hot in the summer, and on one particularly sizzling day, a load of jumpers were all sitting under one wing in the shade waiting for the pilot. One jumper became bored and climbed into the plane. He was never able to adequately explain what possessed him to enter the cockpit and start playing with the controls. When the DC-3 was built, the engineers devised a method of raising the landing gear one wheel at a time so that a smaller motor could be employed. That design detail was to save the life of thirty people. The guy in the cockpit managed to release the landing gear and wheel on the side opposite where the jumpers were sitting, suddenly retracted, slamming the wing to the pavement. The stunned jumpers saw the wing over them rise into the air. Had they been sitting on the opposite side of the plane, they would have been squashed like bugs!

* * *

Our jump planes were workhorses and took a lot of abuse. Occasionally a battery would get weak, and I'd be called upon to "prop it" just like the mechanics did on the old World War One biplanes. On our Cessna 182, I'd pull the propeller until I could feel the cylinder compressing and yell, "Make it hot". The pilot would turn on the key, and I'd make a vigorous pull, stepping out of the way. It frequently took several tries. We had a modified Cessna 195 whose seven cylinder 300 horsepower radial engine had been replaced with a 450 model with nine cylinders. It was a beast. My friend Jim was asked to prop it. I don't know if the ignition was on when it shouldn't have been, or Jim just got distracted, but the engine fired up before Jim got clear. He was unscathed, but the spinning prop picked his watch right off his wrist! We were both done with propping jump planes after that.

* * *

For a brief time, we jumped from a sod farm in the town of Brandon, FL. The takeoff area had a large creek with a steep bank running across our "runway". The plane couldn't get enough airspeed to take off by the time we crossed the ditch, so the pilot would yank back hard on the yoke and drop the flaps causing the plane to lift just enough to clear the gap and then settle back on the ground on the opposite side. He'd then continue accelerating to the point where he could do a normal takeoff. This arrangement worked well until one day, the pilot tried his hopping maneuver too soon, and the plane settled back down without entirely clearing the ditch. The door was off the plane, and I was sitting on the plane's floor, transfixed by the sight of the creek's bank approaching.

We slammed into the bank, and dirt flew into the cockpit, but somehow the plane managed to bounce out of it. We continued rolling and lifted into the air. Nobody said anything. We just breathed a long collective sigh of relief. At about 5000 feet, I looked west toward Tampa and saw an approaching jetliner heading right at us. I yelled at our pilot, who yanked the plane hard left, and I saw the passenger jet simultaneously crank a very hard turn and pass to our rear without a lot of clearance. I hope they hadn't started serving coffee on that flight. At that point, I had had enough and climbed out on the strut. The pilot yelled, "We're nowhere near the exit point!" and I shouted back, "I'll take my chances!" as I dropped away. That trip to altitude had disaster written all over it, and a lengthy hike on the ground was a small price to pay.

* * *

Indiantown in south Florida was the home of a drop zone owned by one of the legends of the sport. Paul Poppenhager had more jumps than just about anyone in the world. I watched his 5000th jump many years ago and wasn't sure he would open in time. That dude smoked it low before the canopy appeared.

He had a Cessna 150 designed for only two occupants but would use it to drop two jumpers. After the jump run one day, the Cessna descended, and the engine quit. I don't recall who was flying it that day, but there wasn't altitude left to make it back to the grass airstrip. We watched as the plane settled into an orange grove. The pilot held it aloft as long as he could and then yanked up on the yoke causing it to stall. It dropped neatly onto an orange tree, and he escaped unhurt. A large team of volunteers extricated the plane and hauled it back to the airport. Whatever caused the engine to quit was quickly fixed, and the plane was back in service.

You know those swiveling drink holders that can be found on boats? This drop zone was the only place I ever saw one mounted on the instrument panel of a jump plane. The pilot could do a complete barrel roll and not spill his beer.

* * *

In the early seventies, several jumpers from the Zephyrhills Skydiving Center in Florida decided to compete in a meet in Hammond, Louisiana. It coincided with Mardi Gras, and we had no trouble finding a dozen passengers to fill up the 1937 Lockheed Electra L10-E, which, coincidentally, was the twin of the one Amelia Earhart was flying when she disappeared. We had an uneventful outbound flight. The competition was cancelled because of a low deck of stratus clouds, so we headed into New Orleans early. It was on that trip that we discovered Boone's Farm Apple wine. Unfortunately, our pilot also made the discovery. He was so hung over the next morning that two of our members bought tickets on a commercial flight to get home. The rest of us couldn't afford that option, so we decided to take our chances.

The cloud ceiling was still solid at 1200 feet over the entire southeast United States when we lifted off. The plane had no facilities, and no one thought to bring anything to eat or drink. I was so dehydrated that I was licking moisture that had gathered on the small port window. Up in the cockpit, the bleary-eyed pilot had climbed the plane through the same solid deck of low-hanging stratus that refused to go away. He was flying on instruments when the artificial horizon went on the fritz. Keeping a plane level is a fairly simple matter when you can view the horizon, but it's a totally different matter when you're flying blind. Sitting on the floor next to the pilot was a half-empty bottle of Boone's Farm. He grabbed it and drew a horizontal line at the level of the wine with a grease pencil. He then duct taped the bottle to the instrument panel. Ingenious! Now, when the plane rolled on its axis, the wine in the bottle tilted and he could bring it back level.

His other navigation instruments also shut down, so we found ourselves flying over the Gulf of Mexico in zero visibility, relying on a wine bottle and a magnetic compass. At this point, pretty much everyone on the plane needed to use a bathroom, and we were clueless about where we were. We continued heading roughly east southeast, trusting luck we'd break out of the cloud cover somewhere near Tampa.

The cloud cover abruptly ended when we crossed the coastline. Unfortunately, nothing looked familiar. Despite protestations from cross-legged passengers in serious need of facilities, our pilot had no place to set the plane down. He continued east until he spotted a good-sized airport. Mercifully, he landed, and we raced to the airport's small bathroom. When we emerged, we discovered we were in Gainesville, 120 miles north of Tampa. From there, the flight to our final destination was uneventful, but I've always wondered how close we came to being one of the never solved missing plane stories.

Sometime later, that same pilot showed up at the drop zone on four successive weeks driving a Corvette, a Jaguar, a Porsche, and a Maserati. I asked him about his good fortune, and he told me he had been contracted to fly tropical fish to the states from South America. I don't think anyone believed him. Last I heard, he was in prison in North Carolina.

That same Lockheed figured into another memorable jump at Zephyrhills, Florida, in the early seventies. Exiting from 12,500 feet provided enough altitude to allow a full minute of falling at terminal velocity before opening. I proposed that we open our chutes immediately up exiting – more than two miles up! Calculating the exit point would be challenging, and any miscalculation could result in us landing WAY out in the boonies. Florida has some seriously desolate boonies. I was the morning meteorologist at Tampa's NBC-TV station, so I called a coworker and asked what wind speed and direction were reported on the 700 millibar chart, which is roughly 10,000 feet and considered the top of the lower atmosphere. With that limited knowledge, ten of us ascended to 12,500 feet. I suppose we should have let the FAA know what we were doing because we were certainly high enough to be in the paths of commercial jets.

The winds aloft were quite strong, and it was literally a "leap of faith" when I selected the exit point several miles from the drop zone and stepped out, and executed a "hop and pop". The other nine dutifully followed my lead, and we looked like a troop of paratroopers. We were all jumping round parachutes – certainly steerable but far less maneuverable than today's ram air square chutes. The drop zone was WAY off in the distance, and I began "running" – riding the wind toward the target. The ground speed was very fast, and I switched to "holding" – facing into the wind to slow our speed. I repeated that pattern over the next ten minutes – guessing and adjusting. Remarkably, all ten of us landed on or near our 30-meter pea gravel target.

With today's high-performance canopies, "Spotting" is a lost art, but on that day, I'm happy to report my choice of exit points was "spot on".

* * *

Over the course of my jumping career, I made a lot of demo jumps. I was happy to do it for the free jump, but once in a while, I got paid. For a while, Connie and I, along with our friends Ski and Donna, jumped into nearby Cypress Gardens for four shows a day. We had to land in the lake each time. Between shows, we needed to air out our canopies to dry them. Since I was the only FAA-licensed rigger, I had to repack all four reserves each time. I'd get the job done just in time for our next show. It was fun performing before an appreciative crowd, and I liked to show off. On the first jump, I waited until I was about a hundred feet up and disconnected the leg straps of my harness, sitting only in the harness' saddle. At fifty feet, I grabbed one of the leg straps with one hand and did a front somersault dangling below the descending chute. Just above the water, I swung my legs up, letting go and doing a flyaway dismount and entering the water. The crowd loved it, but one of the waterski performers told me afterwards that there are "cypress knees" just below the water surface, and I might have been impaled on one. I abandoned that stunt and contemplated the next one.

Cypress Gardens had a large overhanging sun shade over the crowd, and a pier stuck out in the water. The narrator would stand at the end of the pier describing the action. This was in the days before the more maneuverable square parachutes, but I figured I had enough control of my canopy to fly under the overhang and land on the pier. The narrator's eyes got real big as I swooped in. He realized what I was doing and stepped out of the way as I touched down on the pier, taking several big strides and finishing up on a short section of beach in front of the grandstand. The crowd went wild, and both my main and reserve chutes stayed dry! I was discouraged from repeating that particular stunt.

In May of 1975, I got to make a jump with Richard Bach, author of Jonathan Livingston Seagull. Richard was new to the sport. I "flew" in freefall and made contact with him. He told me I was the first person he had ever seen in freefall. When I completed my 2000th jump, he presented me with the Diamond Wings award from the US Parachute Association. Sadly, the roll of photos shot that day was lost by the film processor.

Skydiving was a huge part of my life, and I'll always treasure the memories.

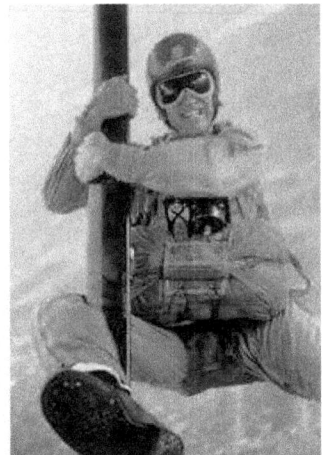

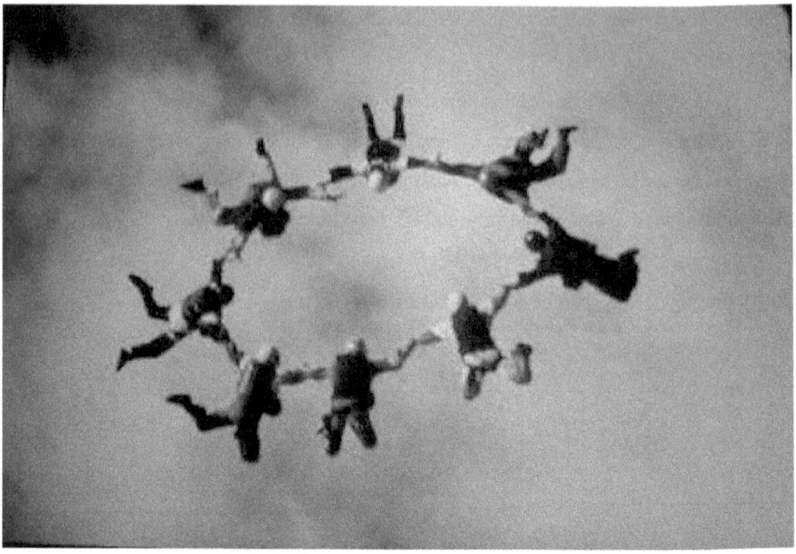

Running boards

Are you old enough to remember running boards on pickup trucks? In the old movies, the G-men would stand on them in high-speed pursuit while firing their guns at the bad guys. I had always wanted to stand on one while a truck was rolling. I got my wish. A fellow student named John at the University of South Florida was the jump pilot for our skydiving club and, during a semester break, invited me to visit his home in Vero Beach, FL. When in high school, John and twelve of his friends each contributed $100 to buy an Aeronca Champ two-seat airplane for $1300. John and I spent some time logging hours in the plane.

Back to running boards. John's father (who was the Chief of Police) owned a vintage pickup. I convinced John to cruise along a back country road while I climbed out of the truck. Adult beverages might have influenced my thought processes. Anyhow things were going swimmingly until I had the brilliant idea that I would pretend to be a trophy deer that some hunter had lashed to the front fender. As we headed down the road, I climbed up on the fender and stretched out. John hit a sizeable bump, and I grabbed the radio antenna that broke off in my hand. Time kind of slowed down as I found myself sliding off the side of the truck at 40+ mph. Instincts took over, and I executed a perfect PLF (Parachute Landing Fall). Actually, I think I did two or three PLFs in rapid succession before coming to an ignominious stop. John got the truck stopped and ran back to see if I was still alive. Other than a bruised hip, the only other casualty was my pair of Levis. My bout with "road rash" resulted in the disappearance of a big strip of denim material. I was done with riding running boards.

Caving

While at the University of Florida, I heard of the existence of the nearby Warren Cave, which at four miles long, was the largest dry cave in the state. With two dorm mates, I decided to explore it one afternoon. Until 1959 the existence of several narrow passages called "squeezes" wasn't known. They led deep into the cave. We were determined to find one and were successful. As the skinniest explorer, I opted to go first. This particular squeeze was very narrow and had a pronounced sideways slant straightening out at the very bottom. With a lot of effort, I managed to get through to a fairly large area but to say it was a tight fit going through is an understatement. I can't recall the name of the second guy to try, but he wasn't quite as skinny. Negotiating the tight passage feet first, one of his feet got jammed in the narrow crack just before the open area. Despite his best efforts, the foot was twisted at an awkward angle and wasn't going anywhere. It was too early for panic, but my flashlight's battery wasn't going to last forever. We debated having the third guy drive back to Gainesville for help but realized I had the car keys in my pocket. After an hour of futile efforts, the poor guy was hyperventilating, and I had the bone-chilling thought that I would spend what was left of my life in a stygian cavern.

I could reach his foot and had been trying to dislodge it without success. At one point, he actually suggested that I break his foot to free it! Clearly, he was desperate. Even if it would work, I didn't have the strength to do something like that, but I persisted in my efforts. Somehow I got his foot out of the shoe! Half an hour later, he dragged himself back through the squeeze, and I followed him. It was a sobering experience and not one I'd care to repeat. Fifty years later, that shoe is probably still stuck in that squeeze.

Water Skiing

Dad had always wanted a fishing boat. When we lived at Westover AFB in western Massachusetts, he bought a 15' Crosby with an Evinrude 45 hp outboard motor. My siblings and I pleaded with him to purchase water skis, and he agreed. With our newly acquired gear, we headed for the nearby Connecticut River, where we rapidly mastered slalom skiing, 360 banana tricksters, and saucer riding. When a friend appeared with a more powerful boat, we built pyramids.

I wanted to learn barefoot skiing, and Dad was very tolerant of my repeated failures until I finally got the hang of it. Fortunately, I have big feet because the boat's top speed was only 28 mph. The water would spray over my head so much I couldn't see where I was going, but I was able to stay up for lengthy periods of time.

We moved to the Washington, DC area, and Dad purchased land along the Shenandoah River in Front Royal, VA. He, my brother, and my brother's friend Doug built an A-Frame cabin. We upgraded to a 16-foot Lone Star boat with a bigger engine and then a 17-foot Crestliner with an inboard/outboard 90 hp Chrysler engine. At last, we could go fast. This boat could hit 40 mph. Suddenly I was able to plane instead of plow while barefooting, and I could even do it on one foot!

While a student at the University of Florida, I did some water ski jumping. As we grew older, we sold the cabin and the boat and moved on to other things. Dad never did get to fish from any of the boats.

* * *

I was a varsity diver for the University of Florida. Our team hadn't lost a dual meet in ten years, but that streak ended at the University of Alabama. My poor vision wasn't a problem diving at outdoor pools, but I struggled to try to figure out when to stop flipping on dives at indoor pools.

On one dive, I managed a painful belly flop. The resulting low score factored into our loss. At dinner that night, the wait staff seated me with our coach Bill Harlan. Coach Harlan could have taken the high road and not blamed me for the loss, but that wasn't his style. He kept saying, "You only had to stick one dive, Moore!" repeatedly. I lost my appetite. At the SEC Conference Meet later that year, we won the title, and Coach Harlan walked up to the Alabama coach and threw an Alabama Football player's raincoat at him. He was never under consideration for any sort of Sportsmanship Award. At a dual meet at East Carolina College (now a University), we were so dominant that he decided to enter the four divers as the 200-meter freestyle relay team as an insult to ECC.

I was NOT a strong swimmer. The first 25 meters out went fine, but I decided to try one of those flip turns. I rotated, pressed my feet to the fall, and pushed off. Unfortunately my trajectory was angled down, and I chose that inopportune time to take a deep breath! I shot to the surface, gasping and coughing, only to hear Coach Harlan screaming at me to try to catch up.

I got even. A short time later, we were competing in our last meet of the year, and the hotel lobby had a carpet that produced a strong static charge. I rubbed my leather shoes against the carpet and took out a brass key. I walked up to Coach and aimed for the back of his crewcut hair. The blue spark, accompanied by a loud cracking sound, jumped across and scared even me. I thought he was going into coronary arrest. It was probably a good thing that I dropped out of school and enlisted.

* * *

A few years later, I was competing at the National Parachuting Championships in Marana, Arizona, and headed for the pool after the day's competition ended.

I won a beer there by betting I could walk across the bottom of the pool. I was in very good shape then. I took in as much air as I could and attempted to float. My feet dropped down, and then my body descended until I touched the bottom of the pool. I intentionally fell over and lay flat out before standing up and walking the length of the pool. Back then, I could hold my breath for over two minutes. I won a LOT of beers in my early years.

The pool had a pretty good quality diving board, and I was still able to do pretty much all the dives I had performed in college.

There was an uptight female lifeguard who was constantly blowing her whistle at the least infraction. When I knew for sure she was watching, I got off my chair, walked to the board, climbed the ladder, and stepped forward. I hit my hurdle step, bounced once, and launched upwards, doing a gainer in the tuck position. My feet landed back on the board, and I reversed everything: hurdle step, backwards walk, descended the ladder, and walked backwards to my chair, where I sat down. The whistle didn't stop until she screamed that I was kicked out of the pool! It was worth it.

My Brief Pugilistic Career

Search your memory banks for the hottest days you've ever experienced. Then add a few more degrees and suffocating humidity, and you have Fort Benning, Georgia, in July and August of 1968. This military complex rivals a blast furnace at full output. I was attending Army Jump School along with 1500 other would-be paratroopers. We were housed in un-air conditioned barracks that predated WWII. Lower bunks on the first floor were highly prized as the sweltering heat enveloped you at night. More than fifty years later, I still recall stretching out in my bunk, eyeing pools of sweat forming on my skin. As an Air Force Sergeant, I was fond of my creature comforts and not at all enthused about the treatment we got. I guess we were supposed to learn discipline, and the Army determined that harassment and poor sleeping conditions would produce dedicated warriors who would never question an order. This lasted for three weeks: Ground Week, Tower Week, and Jump Week. I was looking forward to the latter, but the first two were less appealing than a root canal.

To make matters worse, the instructors kept devising devious means to remind us who was in charge. I'll never forget blaring loudspeakers at 3 am screaming, "Fall out for linen exchange". Our sadistic Company Commander ordered us to participate in a boxing match. I have no idea how they figured this would make us better paratroopers, especially in my case. Until a growth spurt at age sixteen, I was usually the shortest kid in school (boy or girl) and was pretty good at avoiding physical altercations through diplomacy. There was no way to talk my way out of it at Fort Benning in the summer of 1968.

I was given an ill-fitting, padded helmet and an oversized pair of gloves that looked like they'd be appropriate for a circus clown. As soon as a grinning NCO had brutally tightened my gloves, I developed an itch on my cheek. It's physically impossible to scratch an itch while dressed like Ronald McDonald. That issue suddenly became less pressing as I found myself in the ring looking up at another student from our barracks.

I don't know if he was a former Golden Gloves Champion, but he sure looked the part. He was about 6' 12" and had a reach that made me wonder if his knuckles dragged on the ground when he walked. He had a sadistic grin with a streak of drool dripping down one side of his mouth. A fly sat perched on the tip of his nose, and he didn't even notice. Right then and there, I knew he was in some kind of zone that would only end when he beat something senseless. He danced from foot to foot with an animal's grace and began making short jabs at an imaginary punching bag. I took this all in with my "deer in the headlights" countenance and tried to figure out the best way to survive this. When the gong sounded, I covered my face with both hands while he slammed his fist into my vulnerable tummy. I gasped and dropped my hands, only to have my head rocked back by another swing.

The crowd screamed their approval, and I envisioned a Gladiator facing a hungry lion at Rome's Coliseum. Time dragged to a halt as I flailed my hands in a windmill fashion but encountered nothing but air. Suddenly he grabbed me and put me in a bear hug. He smelled like a goat and was actually laughing! We were both so sweaty I could slip out of his grasp.

Everything after that was a blur. I rapidly became adept at hitting his hands with my face. My legs refused to support me, and I sank to the mat. I think he scored the quickest TKO in history as I hit the mat and refused to get up. As I lay there dazed, my tongue took inventory of my teeth and tasted blood. The referee counted from one to ten way too slowly. I was helped to my feet and sent to the infirmary.

I had assumed my experience with fisticuffs was over, but that was not to be.

Fast forward a couple of years. Connie and I had been competing in a skydiving meet at the Kissimmee, Florida airport and decided to join a few other jumpers for a beer that evening. It happened to be the weekend of the annual Silver Spurs Rodeo. When we got to Roscoe's Bar, our olfactory senses were assaulted by a perpetual nicotine cloud while a mediocre band was making up in volume what they lacked in skill. A contingent of dime store cowpokes in ten-gallon hats was unsuccessfully trying a line dance.

We went inside anyhow and really stood out as we were wearing baseball hats. After several beers, Connie and I were dancing when she suddenly grabbed me by the arm and took me outside. I had been looking at the band and missed what had happened. A drunk cowboy had flipped the brim of her hat up. She responded by giving him the finger, whereupon he pretended to take a swing at her and accidentally did make contact. I had missed all that. She said she was okay but just wanted to go home. I told her I had left something in the bar and would be right back. I found the guy and, in a reasonable tone, told him he needed to go outside and apologize. His response was a crude expletive, and the adult beverages in me kicked in. Just like at Fort Benning, time slowed down as I swung a haymaker at his face. The blow knocked him right into the band's drum set. Incredibly, the drummer never missed a beat and looked like this was nothing new.

Fearing my opponent would leap up and start pummeling me, I ignored the Marquis of Queensbury Rules and dove on top of him, continuing to beat him as hard as I could. I repeatedly slammed my fist into his face and only stopped when it started getting bloody. I looked up to see that I was surrounded by dozens of ten-gallon hats. Fearing extreme bodily harm at the hands of the mob, I leaped to my feet and bolted through the crowd. The doorway was blocked, so I escaped through an open window yelling at Connie to get in our decrepit '66 VW. Normally a slow starting machine, it happily fired up, and we took off as fast as it would go. I expected a posse of pickups to chase us, but we mercifully escaped unscathed.

The next day I learned that the guy's wife saved us. She stopped our would-be pursuers by telling them that her imbecilic husband deserved what he got.

I have permanently retired from fisticuffs with a 1-1 record.

Flying

After making a bunch of jumps, I decided I'd like to learn to fly. I signed up with a flight school based out of Tampa International Airport and got the needed medical clearance to become a student pilot. Most flight schools have easy-to-fly planes like the Cessna 150. Our school had Grumman American Yankees, two-seater low-wing planes that had to be landed at a higher speed. Nevertheless, I started lessons and made pretty good progress, eventually soloing. It was pretty gutsy to solo the first time at an airport as busy as Tampa's.

Part of the training involved a check flight with a different instructor. When we were about 2000 feet, he pointed out something on the horizon to distract me and pulled the mixture control killing the engine. I guess he wanted to see how much I'd freak out. I instantly spotted what he had done and slapped the control back in, saying, "Don't do that!" The engine immediately fired up, and he grinned. I guess I passed whatever test he had devised.

One of my skydiving buddies was an Air Traffic Controller at Tampa International. I was flying solo one day and wanted to practice touch-and-goes and asked the tower for permission. The response I got (I recognized my friend's voice) was, "79 Lima, winds are 09 at 20 gusting to 30. Land on runway 01 Right." This meant landing a very lightweight plane in a very strong crosswind. Not only did this exceed my ability, but the plane wasn't well suited for it. I didn't even have enough sense to stay on the upwind side of the runway. All was going well until I chopped the power, and the plane started drifting hard to the left. I stomped on the right rudder and held my breath as I flew in a crablike fashion. The plane got lower, and I straightened it out just before touchdown on the far left side of the runway. I did some impressive skidding and fishtailing. The voice on the radio was laughing and said, "79 Lima, if you can ever get that thing under control, there's a 737 on your tail. I slammed the throttle forward and took off, responding, "79 Lima requests a go around and a landing on Runway 10". I got a chuckle and a "10-4" from my friend and made an uneventful landing directly into the wind. My instructor was on the ground listening to all this on the radio and was pissed at the Air Traffic Controller. I talked him out of filing a complaint.

Shortly after that, we learned that we were expecting our first child, and my flying career ended with only 30 hours of flight time.

Gordo

I competed in a lot of Florida Parachute Council competitions around the state and knew most of the judges. At one event near Cape Canaveral at Rockledge, there was a new person judging accuracy. He looked somewhat familiar, but I couldn't place him. Later at the Top Hat Bar and Grill, I spotted him at the bar and asked one of the other jumpers who he was. It was Gordon "Gordo" Cooper, one of the original Mercury astronauts who later flew a Gemini mission. I introduced myself and then unwisely told him a joke about Polish astronauts. He stared at me with a steely gaze and said, "I've heard it. And my wife's Polish". I froze and was horrified at my gaffe when he burst out laughing and said, "Just kidding". Relief swept over me.

I had one of those vans with carpeted walls back then, and it had a black and white TV with a tiny screen. A show about Acapulco cliff diving was being broadcast, and Gordon wanted to watch it as he had done some cliff diving himself. A case of beer appeared from somewhere, and we talked for a few hours. He told one story after another, but the only one I remember was him saying he had been the navigator on a boat in a Miami to Nassau race, and they had gotten lost. He was a fascinating individual. I was teaching Earth/Sun Relationships as part of the Geography curriculum at the University of South Florida, and I asked if he would be a guest lecturer in a future class. He agreed, but it never came to pass.

Never got shot from a cannon

Kids dream of running away and joining the circus. In 1968 it wasn't practical because I was still in the Air Force. One of the great disappointments of my life was missing out on a chance to get shot from the Zacchini Brother's cannon. My friend Bill Ottley introduced me to George and Nancy Rucavado in nearby St. Petersburg. They had appeared at the 1964 New York World's Fair as Puffed Rice and Puffed Wheat being shot from twin cannons. Obviously, Quaker Oats was their sponsor. I met them for dinner on a couple of occasions and watched home movies of Hugo Zacchini being shot over a couple of Ferris Wheels in Las Vegas.

The cannon was controlled with compressed air. The explosion and smoke were just for effect. Getting the correct trajectory and speed was difficult. One time in Texas, Nancy's feet hit Wallenda's highwire, and she barely landed in the net, breaking her leg. George died in a tragic accident in South America, falling through a false ceiling while rigging a show for Disney. Adventures like these can be risky, but the experience is worth it.

I really wanted to enjoy the thrill of being shot from a cannon. Sadly Hugo nixed my request.

Air Force

After two years at the University of Florida (named America's #1 Party School that year by Playboy Magazine), I had managed to maintain a 1.9 GPA, and the draft board was breathing down my neck (there was a war on, you know – it was in all the papers). I could have returned to school under the auspices of academic probation, but I made the decision to enlist in the Air Force. A few weeks before induction, a high school friend and I attended our first outdoor concert. It was held at DC Stadium on August 15, 1966, and the opening act was Cyrkle, who had hits with Red Rubber Ball and Turn Down Day. After they finished their set, two limousines drove onto the field, and we watched John, Paul, George, and Ringo emerge and climb the raised platform at second base.

We were sitting in the second row behind the third base dugout. Cops were lined up in front of the dugout and surrounding the platform the Beatles were playing on. A guy sitting in front of us jumped up, hopped over the dugout and the cops, and sprinted for second base. All the cops there were watching the Beatles and didn't see him coming. He squeezed between them and jumped up on the stage, shaking hands with all four Beatles (who never missed a beat) before diving into the arms of a zillion cops. They hauled him off as we applauded. The next day the newspapers reported that he had been released on his own recognizance, and I thought to myself, "Why didn't I think of that?!"

In any event, we got a chance to enjoy the music of the greatest band in history. It was a nice sendoff to my military career.

Induction into the Air Force was at Ft. Holabird, MD. Dad took half a day off from work and drove me to Baltimore to report for duty. I got there, signed in along with about fifty other inductees, and we were told to take off our clothes. A flight surgeon lined us up buck naked in two rows and said, "All right, bend over and spread your cheeks!" Honest to God, the guy next to me looked confused, bent over, inserted one finger in each side of his mouth, and tugged.

We got dressed and filled out countless forms. Then we experienced the first of many examples of military bureaucracy – they had scheduled too many people for induction that day and told us to go home and come back tomorrow. I hitchhiked back home and walked in the door. Dad must have been thinking, "The kid's been in the service less than one day, and he's already gone AWOL"! I explained what happened, and the next day he drove me back up there.

Lackland AFB near San Antonio in 1966 was referred to as The Armpit of the Air Force. Thousands of enlistees spent six weeks in blistering Texas heat at basic training before being shuffled off to tech schools. I learned that the war effort depended on us knowing that handkerchiefs had to be folded bottom-to-top, right-to-left, right-to-left, and bottom-to-top. As a side note, I received my draft notice two weeks *after* I got to Lackland.

You know how strange coincidences pop up in your lives? In my junior year of high school, I was a diver on the swimming team of Chicopee Comprehensive HS in western Massachusetts. One of my teammates was named Stuart Chalmers. Four years later, I was sitting on the curb outside North Hall at the University of Florida, and Stuart walked by. We chatted for a bit about what a small world it is. Now here I was at Lackland AFB learning how to fire M1 rifles (they still trained with them in 1966), and I discovered the guy next to me on the firing line was Stu! About ten years later, I was the jumpmaster of a class of first-time skydivers in Zephyrhills, FL, and one of my students told me her boss in St. Petersburg said to say hi. You guessed it – Stu Chalmers.

The Air Force was supposed to send me to Radio Intercept Analysis School, but they had booked too many students. As a result, I was assigned to "casual status" at Lackland, where I worked in a freezing cold meat locker. One week stretched to two, then three, and the powers-that-be decided they would send me to Air Police school instead. I had vowed that I would never use Dad's influence to help in my Air Force career, but visions of saluting car bumpers in Minot, ND, were the deciding factor. Dad was a retired Air Force Colonel and made a call. As a result, I was sent to Chanute Field in Rantoul, IL, to learn to be a weather observer.

Considering my poor academic performance at the University of Florida (where my English teacher dubbed me Conan the Grammarian), I surprised myself by graduating first in my class at Weather School. With that honor came the choice of assignments. I selected the 1st Weather Squadron at MacDill AFB in Tampa. The spot was for an Airborne Weather Observer, and it meant I'd make military jumps in advance of Airborne assaults. Here's what the job entailed: Let's say the United States needed to quell an uprising somewhere in Africa and was going to drop hundreds of paratroopers from the 82nd or the 101st Airborne Divisions to accomplish the objective. The role of the Airborne Weather Observer was to jump in the night before by helicopter wearing a vest containing several 10-gram helium bottles and a device called a Theodolite which functioned like a small telescope. The helium was used to fill pilot balloons that had water-activated battery-powered lights. Once launched, the Observer would note the direction the balloon flew and determine the winds aloft. As the C-130s and C-141s approached the drop zone, panels forming a large orange T would be set out, and smoke bombs deployed to indicate the exit point. Any screwups and the entire contingent of jumpers could land in the boonies and render the operation a disaster. One Observer was calling out the coordinates of the balloon during a training exercise when his co-worker noticed that he was looking in the wrong direction. The balloon's lights were still visible, and the guy was following a satellite! Disaster averted. Anyhow, I was psyched to be assigned to that job. Unfortunately, somebody at the Pentagon must have noted that a classmate named Sam Roberts was from Tampa, and my duty assignment was switched with his. I was assigned to Hunter Army Air Field near Savannah. Once again, Dad interceded. Sam was so excited to be going home that I didn't want him to be disappointed. I called Dad again, and he said he arranged for both of us to go to Tampa. He said, "There is one condition. Sam's going to have to go to jump school". I said, "I'll break the news to him".

Ironically, Sam went to jump school before I did. I've worn glasses since the fifth grade, and my eyesight was weaker than the limits imposed for jump training. I went to the flight surgeon's office to be tested and wore my contacts *and* my glasses. I whipped off the glasses and passed the test with flying colors. The tester picked up my glasses and noted the extremely thick lenses. He said, "I'm going to need to take a close look at your eyes." It wasn't hard for him to spot the lenses, and I flunked the test. Not only did I have to suffer the indignity of seeing Sam with his brand new jump wings, but he was getting an extra $55 a month for hazardous duty pay! You have to understand that base pay for an Airman Third Class in 1967 was $98 a month. To help pay for my skydiving hobby, I'd sell a pint of blood for $10 every 8 weeks. "You're going to take it easy for the rest of the day, aren't you?" the nurse always said, and I always assured her I would as I drove my 1961 Rambler wagon to the airport with enough money to make four more jumps.

I should point out that my duties as an Airborne Weather Observer were only needed during exercises or actual combat, which never occurred during my enlistment. My other duty was to track weather satellites. I'd get up at 2:15 every morning and drive to the base. I was assigned a Satellite Tracking Unit mounted on a GOAT (Goes Over All Terrain – the Air Force is big on acronyms) trailer hauled by a "Deuce and a Half" two-and-a-half ton truck that contained a large generator. Each morning I'd fire up the generator, activate the tracking antenna, and download the imagery from the ESSA -8 weather satellite. I'd print the result, which showed the cloud cover for our region. In the event we were deployed, I could provide that information to forecasters anywhere in the world. I was assigned not one but *two* C-130s plus their entire crews for each deployment – one for the truck and one for the GOAT. Over a cup of coffee one morning, I met with the loadmaster and, with a pencil on a napkin, showed him how I thought it was possible to load both vehicles in a single C-130. He agreed and tried it successfully. For every deployment after that, I saved the Air Force the cost of flying an entire plane and crew. I should have gotten a bonus.

My Squadron Commander was a Colonel named Sweeney, and he surprised me by submitting my name for the Air Force Academy Prep School. That essentially meant that after two years of college, I would be going back to high school. Completion of the program guarantees you entry into the Academy. When he told me he had done it, he also broke the news to me that I was too old. I recall thinking it was the first time in my life I had ever been too old for anything. I planned on participating in AECP, the Airman's Education and Commission Program. Enrollees were promoted to pay grade E5 and sent back to college to study Meteorology. In order to qualify, you had to have two years of college, a 2.0 GPA, and Math through Calculus.

I enrolled in night school at the University of Tampa and got my GPA high enough. I then took Calculus. It was a struggle. Midway through the course, I was wrestling with Differential Calculus when I was sent to Panama on a military exercise. I recall flying back, sitting in the driver's seat of a deuce and a half truck *inside* a C-130 plane, trying to understand the subject. When I finally got back to MacDill, the class had moved on to Integral Calculus, and I was so lost I gave up. I guess I was never destined to become an officer.

MacDill AFB was part of STRIKE Command (Swift Tactical Reaction In every Known Environment - another acronym!) and had a substantial number of representatives of all branches of the service. The command surgeon was an army colonel named Pratt. Dr. Pratt was in terrific shape and was an avid runner. We became running buddies, and one day, after an eight-mile run, he said, "You really want to go to jump school, don't you?" I told him there was nothing I wanted more, and he signed the waiver allowing me to go. I was ecstatic!

A week before heading for Fort Benning, I competed in the Military National Parachuting Championships at Walkill Strip near West Point, and I finished in third place. Many of the other competitors were instructors at Fort Benning. They took delight in making life miserable for experienced skydivers going through jump training. Here I was with over 400 sport jumps and wearing conspicuous Air Force Sergeant stripes. Even among 1500 students in our class, it was difficult *not* to stand out. Fortunately, I wasn't the only Air Force representative. A couple of hundred cadets from the Air Force Academy were there as well. Curiously, in 1968 West Point wouldn't send their cadets. Someone said they didn't want the cadets to see what the "real" Army was like.

Speaking of the event at West Point, I was climbing to altitude for a competition jump in the Style event, which involves a series of turns and back loops for speed in freefall. Before jumping, we were instructed whether to initiate the series with a left turn or a right turn. For giggles, I used a Magic Marker to mark a large L on the back of my right glove and, conversely, a large R on the back of my left glove. Sitting opposite me on the plane was the commanding officer of the Golden Knights, the US Army Parachute Team. I noticed that he was staring at my gloves the whole way to altitude. We were assigned a series that initiated with a left turn. You guessed it, he turned right!

A month before jump school, I had been in Marana, Arizona, for the National Parachuting Championships. I had been picked to appear in the opening segment of Wide World of Sports' coverage of the event. The show was going to air while I was still at Fort Benning. I really wanted to watch it on the Company's Orderly Room television, but I feared the instructors would figure out who I was. More than thirty years later, my oldest daughter was watching "ESPN Classics" and saw that episode. She recorded it for me. Here's a screen capture from that show. I'm on the left, and my friend Bill Ottley is next to me. Sitting behind the exit door of the Twin Beech aircraft is legendary ABC Broadcaster, Keith Jackson. We were at 2500,' and the camera was rolling. He said, "The 1968 National Parachuting Championships are about to begin. And here they go!" We were nowhere near the proper exit point, but Bill dove out of the plane anyway, followed by me. We had a long walk back.

We went through Ground Week, Tower Week, and finally Jump Week, where we loaded up in Continental Air Command C-119s, "Flying Boxcars," and shuffled out the exit doors at 1100 feet. After five jumps, we were awarded our coveted wings in a ceremony in an auditorium. The student graduating first in the class was given a beautiful bronze statue called the Iron Mike. They called my name, and I went up on the stage to be presented the award from one of the guys I had beaten at West Point. He recognized me and said, "You were in this class???!!" He would have reveled in the chance to see how many times I could push Georgia clay away from my chest. The prospect of doing pushups in August in Georgia was the reason I had tried so hard (and successfully) to remain inconspicuous. I may have been the first member of the Air Force ever to win the award. I still have it today and treasure it.

Top Jumper

"IRON MIKE" is a nine-inch-tall bronze statuette of a paratrooper presented to the Distinguished Honor Graduate of each class at the Army's Airborne Training Center, Fort Benning, Ga. A recent "Mike" winner is Sgt. Pat Moore of U.S. Strike Command's Weather Division. He is believed to be the first STRICOM airman ever to win the honor.

I was now eligible to join the "Airborne Association" and promptly registered for a jump fest at MacDill AFB. It was an accuracy competition using military T10 canopies, which had almost no steerability. My commander, Colonel Sweeney, landed farthest from the target and took a lot of good-natured ribbing. After the competition, we attended a banquet. Colonel Sweeney invited me to sit at his table with others from J-3 Operations. The two-star general heading up J-3 was never very friendly to enlisted personnel, and I approached the table with a lot of trepidation.

Colonel Sweeney held up the small trophy he had won for recording the worst jump of the day. It was a bronze statuette of half a mule. You can guess which half. He said, "Moore, what do you think of this"? As tactfully as I could, I replied, "Sir, it looks like the south end of a northbound mule". The general's wife was on her third martini, and she giggled, "It does? It looks like a horse's ass to me"! The general gave me a look that signaled I should sit still and shut up for the balance of the evening. I was happy to comply.

One of the benefits of being in the service was free jumps from helicopters. The venerable Huey could haul several jumpers to 7500 feet in a comparatively short time. Whenever a dignitary would visit MacDill, the base skydiving club would put on a demonstration of precision skydiving. I didn't always see eye-to-eye with the guy running the club. He wouldn't prevent me from jumping, but I rarely got to meet the dignitary. King Hussein of Jordan was scheduled to visit the base, and the whole base was talking about it. Everyone feared that the Commander of STRIKE Command, General Conway, would show up on a surprise inspection with the Jordanian monarch in tow. Every day at noon, our squadron's four senior NCOs would play cards in the day room. The day King Hussein visited MacDill, I hid outside of the day room and said in a loud voice, "Your Highness, this is our operations center!" Cards flew everywhere, chairs got tipped over, and I figured that I'd better disappear fast.

On the day of the jump, I headed for the flight line where my fellow club members were suiting up. My assignment was to do the max track – I was to "fly" laterally in free fall as far as I could. It's amazing how far a person can move laterally while falling at 120 mph. The only catch was that I was to land in an out-of-the-way place and mind my own business. I was really pissed that I couldn't meet the visiting monarch. King Hussein was a big aviation buff, and I admired the guy. Once I left the helicopter, I began tracking over the middle of the base. At 2500', I opened my parachute and started figuring out where I might land. The base pool looked very inviting below me, but something told me that was a bad idea. I looked west at the hanger row and had an idea.

All the hangers faced away from me and had huge sliding doors. At the center of the bend on the hanger, row stood the hanger housing the KC 135 that General Conway used for travel. My plan was to swoop over the top of the hanger and crank down hard on a toggle turning the parachute so that I would fly right into the hanger. I was chortling with glee thinking about the reaction I'd get out of the aircraft maintenance people inside. All was going according to plan until I realized that the hanger doors were closed! I yelped and tried to crank a hard turn back the other way.

Unfortunately, I ran out of altitude and smacked into the tarmac at the peak of the canopy's oscillation. To this day, I can feel how hard that impact was. I recall a young lieutenant running up to see if I was still alive, and I bounced to my feet, muttering, "No problem, no problem". I went home to soak in a tub and reflect on what had gone wrong.

* * *

During 1968, I made two trips to Panama. The first was a simulated airborne assault on Rio Hato, and the second was a week at survival school. On the jump, we had suited up several hours earlier at MacDill AFB and had flown our C-130 around Cuba before descending to 1100 feet. A week earlier, we had jumped a C-141 cargo jet, and the procedures were markedly different. On the jet, the designers had configured a giant windscreen to combat the jet blast. The last thing they wanted you to do was to leap out into that blast. You simply stood up, hooked up your static line, shuffled to the door, and dropped. My buddy Sam must have been thinking he was still in the C-141. As I followed him to the door and made my vigorous exit, I observed Sam rolling down the side of the fuselage of the plane. Upon opening my own canopy, I craned my neck to make sure Sam was all right. He wasn't hard to spot. He was the guy with the parachute suspension lines twisted halfway up to the canopy! He spent the next minute unwinding and managed to land without doing any major damage. His watch face was completely covered with camouflage paint from the side of the plane.

My second visit to Panama was for an entirely different purpose. My tour at MacDill had been pretty lengthy, and I figured I'd be getting shipped out to Southeast Asia any day. I volunteered for a job as a Combat Information Monitor on an EC-121, which had been a passenger prop plane called the Constellation in the Fifties. I didn't get the job, but the folks at Langley AFB, VA, who handed out assignments, noted that I had indeed volunteered for Southeast Asia and issued me orders transferring me to Udorn, Thailand. A prerequisite was the completion of the "Tropical Survival and Ethnic Orientation School" based out of Albrook AFB in the Canal Zone. I took a commercial flight to Tecumen Airport in Panama and had to bribe someone to make a phone call to have the base taxi come pick me up.

At the orientation on base, we were told we could wear flight suits but no underwear. Cotton garments would rot after a few days. We were going to be dropped from helicopters just a few feet up and float downriver in our Life Preserver Units. This was to simulate being shot down. Each of us had a few survival rations, matches, a machete, and three gores of a parachute. When we were actually in the Rio Chagres, we were escorted by rifle-carrying Chaco Indians in dugouts in case any crocodiles put in an appearance.

Once we made landfall, we were met by an instructor who gave us a class on do's and don'ts. "If you catch a fish, wrap it up and cook it in this", he said as he held up a giant leaf, "but never in this leaf. It's a deadly poison!" I remember wondering if I was the only one in the class who couldn't tell the difference between the two. He then held up two more seemingly identical leaves. "This one is great as a substitute for toilet paper, while this one will set off a case of butt itch so bad that an Australian guy threw himself off a cliff to end the misery". I figured then, and there I could go a solid week without eating *or* pooping. He then told us to watch for three deadly snakes, the Fer de Lance, the Eyelash Viper, and the Green Vine snake. I was beginning to get seriously creeped out.

Among our classmates were some pilots from the Bolivian Air Force, and they had a knack for catching game. My resolve not to eat anything lasted only about eight hours when one of the pilots caught an iguana. Our survival kit included a pot for boiling, so in went the lizard. People joke about everything tasting like chicken, but this really did.

Included with some of the survival packs was a small "over and under" gun. One of the guys used the shotgun feature to bag a monkey. I was ready to draw the line. "I'm NOT hungry enough to eat a monkey!" I vowed to myself. Like everyone else, I sat and stared as bits of gray meat bobbed to the surface of the boiling pot. Apparently, others were hungrier than I was because someone kicked over the pot and dove for a piece of meat. That was the signal for everyone else to dive in before it was gone. I couldn't help myself as I elbowed someone away to grab an arm and started gnawing on it. Visions of the scene at the beginning of 2001 – A Space Odyssey flashed before my eyes. You know the one with the apes pounding the ground with bones while Strauss' Also Sprach Zarathustra thundered in the background?

With a contented burp, I tried to get to sleep. We had been issued segments of a parachute in order to create a hammock. Two gores formed the sleeping sack itself, and the third was a flap to keep out critters. Ripstop nylon is not the most comfortable sleeping linen, and the stitching holding the gores together formed a tight seam that aligned pretty much perfectly with your butt crack. It was like trying to sleep with a giant wedgie. To top it off, it began to rain. Actually, it rained pretty much the entire time we were there. During the night, a spider managed to get by my nylon defense and bit me on the back of the neck. By morning it looked as though I had half a baseball lodged under my flight suit's collar.

We were led to a Chaco Indian village to learn a little about their culture. The chief offered us manioc, a pastelike substance made from roots. His wife, who was probably in her seventies, sat next to him topless. A Specialist Fourth Class sat next to me, and he couldn't stop staring at the woman. Finally, he exclaimed, "Gee, it's just like National Geographic!" We all laughed and, through an interpreter, the chief wanted to know what was so funny. Someone came up with a lame explanation that I'm sure he didn't swallow.

The last part of our training was for us to be set loose with the objective of finding our way back. We had no means of navigation and had to find our own food. We were grouped in teams of three, and I promptly nicknamed my companions Curly and Larry. Before we had been out an hour, I heard a noise overhead, and Panama's clumsiest Green Vine Snake fell out of a tree and landed directly in front of me. I think it was as scared as I was because it took off before I had time to react. The experience did slow us down a bit and made us more cautious. At the time we were dropped off, we were each given a floppy hat. We'd get major points if we made it back with the hat. Of course, our primary objective was to make it back, *period*. The old chief was a master tracker, and the Air Force paid him a dollar for every hat *he* brought back. Remember, the course was supposed to teach us to escape and evade. Our evasion skills were on a par with the Three Stooges, but I figured I'd keep my hat inside my flight suit so I could at least keep it when he eventually found us.

Just when we despaired of ever finding civilization again, the chief leaped out from behind a tree that couldn't have been more than four inches in diameter. He spoke the only English word he knew: "Hat!" The others surrendered theirs, but I employed sign language pointing back down the trail, trying to tell him that I had lost it. He slowly walked up and inserted the tip of his machete in my flight suit zipper and repeated, "Hat!" I nodded slowly and extricated it handing it over, hoping he had a good sense of humor. One of my traveling companions tried to mime the question, "Where do we go from here?" but he just smiled and was gone.

We eventually found our way back to the staging area and were transported back to Albrook AFB. I had not shaved or bathed in a week, and I was in need of some real food. I walked into the shower room at the barracks and turned the shower on the full force while I still wore the flight suit. I slowly took off my boots and peeled off the flight suit throwing it in a trash can. I can't recall ever enjoying a shower more. When I finally dried off and got dressed, the chow hall was closed. I knew I could get popcorn at the base movie theater. When I saw the movie title, "The Fearless Vampire Killers", I almost changed my mind, but hunger won out. It turns out the movie was a spoof starring Roman Polanski and Sharon Tate. This was just a year before she was murdered by the Charles Manson family.

The next day I flew back to MacDill on a commercial flight. Sitting next to me was a woman who appeared to be about eleven months pregnant. Judging by her reaction to turbulence, it was pretty clear that she had never flown before. Even I was getting concerned when I could see the wing tips bouncing up and down. I didn't know how to say "cross your legs" in Spanish, but she managed to make it to Miami without incident.

When I got back to MacDill, my commanding officer called me in and asked me if I wanted to go to Thailand. To be honest, I was kind of looking forward to it, but it would have meant leaving my girlfriend and giving up skydiving for a year. I was puzzled at his question and asked him why he was asking. He said that Joint Task Force 11, of which I was a member, could not function without an Airborne Weather Observer. Sam was assigned to JTF 7, and I was the designated Observer for my team. What he was saying was that a thousand troops would be rendered ineffective if I didn't stay on the job. How could I argue with logic like that? When I didn't respond right away, he said, "Don't sell your car". Within a few days, my orders were rescinded. I happily finished my tour at MacDill. In the midst of the tumultuous war years, I managed to spend an entire tour of duty in Tampa, FL.

I had been scheduled for temporary duty in South Korea, but another airman went in my place, and I'm glad he did. The C-130's crew was composed of practical jokers. The turbulence on the flight over was horrible, and no one wanted to be the first to "toss their cookies". Everyone sat on bench seats along the side of the fuselage, staring at someone across the plane. While unobserved, the crew chief heated up a can of vegetable soup and poured it into a barf bag before sitting down at the front end of one of the rows. He then turned violently to his side and pretended to throw up in the bag. Even then, everyone was able to keep their lunches down. He wiped his lips, sealed the bag, and instructed the guy next to him to pass it down to the loadmaster at the other end of the row of seats. Each person, in turn, passed the warm lumpy bag to the next person until the loadmaster took it, opened it, and started drinking the contents. That did it! People everywhere started upchucking! The crew did have to clean up the mess, but it was definitely a classic prank.

* * *

About a year before my discharge, I got married to another skydiver named Connie Besonen. She wasn't yet 21, so we had to get her parents' permission to wed. We had a $70-a-month apartment near the airbase that prohibited pets. Showing remarkable restraint, she limited herself to five cats. Each was named for parts of a parachute: Bungie, Grommet, Toggles, Twill, and I forget the other one. Her job as a claim rep for an insurance company called for her to do some day trips around Florida, so she would take our 1970 Maverick and leave me my Honda 125 motorcycle to commute to MacDill Air Force Base.

One morning as I was getting ready to leave for work, I noticed that Grommet, an oversize orange ball of fur, was extremely lethargic. This cat reveled in bouncing off the walls most mornings. When I saw him lying there and refusing to eat, I knew something must be seriously wrong. His eyes were barely open, and he made a soft whimpering noise. Knowing how much he meant to my wife, I was frantic. I knew there was a vet's office a few miles away, so I picked up the limp feline and tried to figure out how I would get him to the vet without a car. We had no carrier, and even if I found something, there was no way to attach it to the motorcycle. I was wearing my 1505s, a short-sleeved Air Force khaki uniform, and figured I could place the cat inside my shirt. I wasn't wearing an undershirt, but Grommet was so inert I figured there was little danger of him scratching me. I walked out to the Honda, keeping a wary eye for the property manager who might wonder how I had managed to acquire such a large "spare tire" in so short a time.

I strapped on my helmet and gave a swift push on the kick starter. The tinny engine came to life, and so did Grommet.

He began to mew loudly, and I got the bike in gear and took off. That's when things really got dicey. Grommet panicked and began doing laps inside my shirt. He would dig his claws into my skin and race clockwise circuits faster than he had ever moved before. I got stopped by a traffic light, and a van pulled up next to me. I was trying to ignore the driver's astonished stare as the lower half of my shirt exhibited more movement than an accomplished belly dancer. Grommet came to a screeching halt, stuck his orange head out the top of my shirt, and then dove back inside, retracing his frantic laps, this time in a counter-clockwise direction. His claws were digging in so deeply that spots of blood were beginning to appear on my shirt. The light was still red, but I couldn't wait any longer. I popped a wheelie and took off at high speed for the vets. I pulled into his parking lot and skidded to a halt.

    I ran inside and dragged the cat out of my shirt. The vet and his assistant dropped what they were doing and immediately began treatment – on me. They gave a cursory look at the cat, pronounced him perfectly normal, and then treated my wounds. Later, as I sat in the waiting room waiting for the pain of the antiseptic to subside, I contemplated how I would get the cat home. Grommet lay sprawled on the floor next to me, purring contentedly. Admittedly my thought processes weren't operating at peak performance, but clearly, there was no solution other than the way I had gotten him to the vets. With great reluctance, I picked him up and placed him back inside my shirt. The return trip was a repeat of the outbound journey. I got home as quickly as I could, treated my new injuries, changed into a clean shirt, and headed for work. Slipping under the Venetian blinds, Grommet, like the Cheshire Cat, sat in the window, grinning at me as I rode off.

Exploding Thanksgiving

By my calculation, I've been married seven decades – at least one day in the sixties, seventies, eighties, nineties, oughts, teens, and twenties. I got married for the first time on November 28, 1969. The day before was Thanksgiving, and we had arranged for both sets of parents to meet at the home of Ron and Daryl Smith, our Best Man and Maid of Honor. Daryl had never cooked a turkey before and wanted it to turn out perfectly. She had heard that turkeys should be stuffed, so she packed it with uncooked rice. I mean, she *really* packed it. The entire cavity was crammed with it. It was cooking in the oven for the prescribed time with a casserole dish of green beans on the shelf below.

We were in the living room playing charades when we heard an explosion. The door to the oven blew open, and the turkey flew out and slammed into the wall of the hallway in full view of our astonished gazes. Upon impact with the wall, it stretched out all appendages and looked eerily like Wiley Coyote. Embedded in the carcass were hundreds of shards of Corning Ware from the casserole dish. Stunned silence ensued, and then, inexorably, the bird started sliding down the wall, eventually flopping over in a misshapen lump on the floor.

Daryl began sobbing, and Ron calmly picked up the phone and ordered pizza.

Despite that ignominious start, my first marriage lasted 24 years.

University of South Florida

Part of my duties while still in the service involved weather satellite tracking. I would get up every morning at 2:30 and drive to the air base, where I would go to my mobile tracking van and download the signal from the ESSA 8 or Nimbus 3 weather satellites. I would then print the picture and plot the coastlines to give to our forecasters. USF's Geography Department had constructed a tracking unit but had no idea how to use it. Dr. Dewey Stowers contacted our weather squadron for guidance, and I was happy to provide it. After I got them up and running, Dr. Stowers asked what I planned to do after discharge. I told him I planned to enroll at USF but hadn't really given thought to a major. He suggested Geography, and I agreed. This time around, I proved to be a much better student and finished my degree in fifteen months. My GPA at USF was 3.5. I was offered a graduate assistantship and jumped at the chance.

This was my first experience teaching, and I loved it. I taught Introductory Physical Geography, and most of my students were upper-level Social Science Education Majors. There were some excellent students, but on more than one occasion, I thought to myself, "these people are going to be teaching *my* kids???" Once a young man said, "Mr. Moore, this is going to seem like a dumb question". I stopped him by saying, "Nonsense. There is no such thing as a dumb question". He then asked his question. After a pause, I replied, "I stand corrected".

I was astonished at how many of these students were clueless about the world around them. One frustrating day, I hollered out, "don't anyone move"! I went back to my office and retrieved mimeographed maps of the 48 contiguous states. I handed them out and said, "Fill in the state names"! The results were not very encouraging. One or two of the military veterans in my class managed to get all the answers right, but the rest were pretty pathetic. One Chemical Engineering senior had already lined up a job in Texas but couldn't find it on the map. The only two states he correctly identified were Florida and Georgia. Another student put "Wisconsin" in seven different states but not in the correct one.

I liked to give bonus points on tests and usually employed a pun as part of the answer. On one occasion, I asked them to "identify the following," and I had drawn 50° Fahrenheit, but it was upside down. A few of the students correctly guessed I was going for "temperature inversion," but I did give the bonus point to one kid who wrote down 10° Celsius upside down. Those erstwhile teachers are probably approaching retirement now. I just hope they never had to teach Geography.

Years later, I encountered someone even more geographically challenged. I owned an employment agency in Springfield, Massachusetts. The interview room had various maps of the world on the walls. One day one of my recruiters came to me and said, "You're going to have to handle this one". He pointed to our receptionist, running her finger up and down the east coast of the United States Map. I said, "Tammy, what are you looking for"? She replied, "France". I asked why and she said, "I'm driving there this weekend". Knowing how her mind worked, I asked, "Which town in France"? She instantly answered with "Montreal". I sighed and then pointed out Montreal. And France on another map. Her mouth formed a perfect circle and issued a soft O. How did we ever get to be leaders of the free world???

Just the other day, I ran into an even more geographically challenged individual. We have an Airbnb apartment attached to our house in Ludlow, Vermont. Recently a New York City resident rented our place for two nights. He didn't show up the first night, and I texted him. He got back to me saying an issue had come up and they would arrive the next day. The following morning as he was leaving, I asked him what had happened. He told me he had plugged our street address and city into his GPS. Unfortunately, he never specified the state, and there are twelve Ludlows in America. They had driven from NYC to Ludlow, Pennsylvania, which is northeast of Pittsburgh! They spent the night at Niagara Falls and drove to Vermont the next day.

Back to USF. Dr. Fuson taught a graduate course in Cultural Geography, and I was enrolled in his class. He was telling us about a colleague named Larry at the University of Florida (in Alachua County) who was an absolute nut about Saudi Arabia. He had based his entire career on the culture found in that part of the world. I piped up, "Sort of a Lawrence of Alachua"? Dr. Fuson began laughing so hard that he had to leave the class. I think he even called his friend in Gainesville to tell him. I've never seen anyone laugh so hard in my life.

Life was pretty good. I was getting paid to teach, and the GI Bill was paying me more than it cost to take classes. A few years later, I used up my final year of GI Bill to attend the University of Baltimore School of Law. Their most distinguished graduate was Spiro Agnew.

Graduate Oral exams in our department involved a test where all the professors would sit around a table, and the graduate candidate would look at a Coast and Geodetic Topographical map. The location had been blacked out, and the student was supposed to study the map and determine what part of the country he or she was looking at. Clues would be elevations, population densities, and a few subtle cultural references.

I sat down and took a look at the map. Dr. Fuson said, "Pat, over the next hour, your task will be to identify what part of the United States is represented on this map". I replied, "It's Louisiana". There was a stunned silence around the room. "How did you know that"? He blurted out. "You left the latitude and longitude lines on the map", I replied. They looked confused, and someone said, "You know latitude and longitude well enough to tell location"? I reminded them that I had plotted satellite maps for more than three years and had hand drawn the coast of the United States along with all the meridians and parallels. Nobody said anything for a minute. Then Dr. Rothwell said, "Okay, assuming you didn't know the location from the coordinates, how would you have determined it? I studied the map for a few seconds and said, "Well, the first place I looked at has the name Parish in it". At that point Dr. Fuson said, "Pat, you passed the test".

I completed my Masters in December of 1972, and my wife and I had planned to take some time off to travel. I got a call from USF telling me that Dr. Rothwell had fallen into the tub and would be unable to teach the winter semester. I signed on as an Adjunct Instructor. A few days later, I received a call from Paul Catoe, Chief Meteorologist at Tampa's NBC affiliate, WFLA-TV Channel 8. The morning weatherman had announced on the air that he was quitting to sell life insurance, and Paul wanted to know if I would consider auditioning for the position.

The job called for appearances on air Monday through Friday at 6:30, 7:25, and 8:25 ("cut-ins" during The Today Show) and during the 1:00 pm news show on Mondays and Tuesdays. Since the classes I was teaching were sandwiched around the shows, there was no reason not to give it a try. At the audition taping, I remember asking the cameraman to film me only from the waist up. "Why"? he asked. "I'm so scared, I may wet my pants", I replied. I certainly didn't have a broadcaster's voice, and, as someone pointed out, I had the perfect face for radio, but I still got the job. As it turns out, no one else applied for it. In addition to Paul, the other meteorologist on staff as George Wooten, a retired Chief Warrant Officer with whom I'd served in the Air Force. Paul handled the 6 and 11 pm shows while George did weekends and midday Wednesday through Friday.

It was a fascinating place to work. There was a constant parade of celebrities passing through the station. Many TV stars did local dinner theater in the off-season and would appear on "Today in Florida" to promote their shows. Others were spokespersons for products.

Pat Moore and Weather WFLA TV (NBC) Tampa 1973

During my three-and-a-half year stint, I got to see (and in some cases meet) Vincent Price, Sal Mineo, Gavin MacLeod, Jimmy Dean, Orville Redenbacher, Ginger Rogers, Sky King, Bob Crane, Elke Sommer, and Robert Conrad. Ronald Reagan was making his first run for the presidency and popped his head in the office saying, "Hi guys". On my first day on the job, I was walking down the hall, and a tall, tanned good looking guy was walking toward me. I realized it was legendary motion picture star, Cesar Romero. I said, "You're…" He interrupted me in a stentorian voice and said, "Yes, I *am*!" as he strode past.

I had to get up at 2:30 in the morning to get to the station to prepare for the 6:30 show. Back in those days, we actually drew the fronts, pressure centers, and isobars in Magic Marker on a large colored paper map. The Forecast Board was done with stick-on letters. Toward the end of my employment, we were experimenting with computer-aided graphics, but things were pretty primitive in the early days. We did, however, have the ability to superimpose a photo or video on the blue screen behind the set. We were always warned not to wear anything light blue as it would disappear. Milt Spencer, our venerable sportscaster, forgot and bought a new powder blue sports coat. He was a little late for the opening shot one evening, and viewers at home were treated to his head floating along the Tampa Skyline before hovering over his seat.

I took my job seriously. Even though I had been a weather observer in the service and had taught Weather and Climate, my degree was in Geography. I submitted a tape of one of my shows to the review board of the American Meteorological Society and, in May of 1974, was awarded Seal of Approval #110 for Television Broadcasting.

There's a huge difference between live broadcasting and taped. With the latter, you can always do a retake, but with the former, once it's been aired, you can't take it back. The cardinal rule for bloopers was to never correct yourself. Only half the audience might have caught your gaffe, but if you tried to undo the damage, you were just compounding the problem. I had heard of a TV weatherman who had been giving snowfall reports for the northwest and said, "During the night, Helena got six inches. Helena, *Montana,* that is!"

The first time I was asked to substitute for Paul on the evening news, I was understandably nervous. I would be appearing with different "talent" (a term loosely, and sometimes inaccurately, that applies to anyone before the cameras), and we would be using three cameras instead of the two I was used to. As we were coming out of a commercial, the floor director said, "You are about to be watched by a million people who've never seen you before. Stand by!"

I felt like a Zombie. I forced myself to calm down, and I launched into my presentation. It was going smoothly until I tried to say, "High pressure persists". What came out was "High-Pressure per-pissed". Out of the corner of my eye, I saw sportscaster Mike Dotson laughing and pointing at me. Guys on the set took great pains to trip up their co-workers and make them look silly. Admittedly I became a little flustered and began speeding up my presentation. It's crucial that you adhere to precise timing as commercials begin rolling ten seconds before they actually appear. In this segment, I was to "hand off" directly to Sports. Thirty seconds early, I said, "Now let's turn to Sports with Mike Dotson". Mike clearly wasn't ready and couldn't find his opening story. He tried to ad-lib it. "Jack Nicklaus had his worst round of golf for the season today. He now has a fifty-four whore total of….." Mike froze and put his head right down on the desk in front of him. The show kind of went downhill from there, and it was quite a while before I was invited back to prime time.

George Michele gave a fishing report every day during the mid-day news. It was pretty much the same report every day, but he had a loyal following. George would appear on the set during the opening and then go back to his office until he was scheduled to do his bit. He was notoriously late, and I frequently had to stretch a little as he would come running around the corner to take his seat. One day, he fell asleep in his office and totally missed his segment. The floor director gave me the biggest "stretch" sign I had ever seen. I needed to fill almost five minutes! Unfortunately, I had already finished the forecast. The switchover to Daylight Savings was going to take place that weekend, so I spent a little time talking about that. I then rambled. I told the viewers that daylight savings had been explained to a Seminole chief who replied, "We have a member of our tribe who does something similar. He cuts a strip of material off the end of a blanket and sews it on the other end to make it longer". That was lame, I thought. Okay, I was running out of things to say.

I launched into a discussion of weather lore and then explained the differences between Chinook, Bora, and Mistral winds. I gave a short course on Monsoons. The clock never moved so slowly. Mercifully, we finally went to a commercial. I went back to George's office and found him snoring loudly. I then returned to the studio without waking him. The director complimented me on pulling off the "stretch" but he knew I had been struggling. "You know what? Next time, just sit down when you're done, and we'll aim the camera at his empty chair for five minutes".

Most shows went without a hitch, but I had the tape holding the map come loose once. It flopped down over my head and then ripped in half, tumbling to the floor. I finished the broadcast standing next to a blank piece of plywood.

We hired a young graduate from Northwestern University with a degree in Broadcast Communications. It was her first job out of school, and she was brought on board as a floor director for $90 a week. The job entailed sending time signals to the talent. It became quickly apparent that none of her classes had included lessons on how to read a clock. She was clearly chronologically challenged but was still muddling through her first shift when it came time for me to go on. The time commands were pretty simple. Holding up a certain number of fingers told you how many minutes you had remaining. Crossing your arms meant 30 seconds. Twisting one arm signified 15 seconds, and the last ten seconds were counted down one finger at a time.

I had to pace myself pretty accurately so that I finished my segment at the exact second. For example, at three minutes, I would walk from the national map to the local map. At thirty seconds, I would go to the Forecast Board. She held up three fingers, and I remember thinking I had less time remaining. I began to stretch out my presentation a little. A sudden confused look appeared on her countenance, and she made an "erasing" sign. She immediately held up two fingers, and I thought, "Okay – 2 minutes," but she followed that with the thirty-second signal! (She explained later that instead of three minutes, I had two and a half hence the "2-minute" and "30-second" signals). All I knew at the time was that I had only half a minute to cram in five times that much information and started talking like a tobacco auctioneer. The cameraman poked his around the side of the camera and gave me a puzzled look. The young girl suddenly realized what was going on and then gave me a huge stretch sign. I shifted into slow motion, and even my voice took on a languid droning tone. Mercifully, my segment ended. She didn't return the next day, and I suspect she decided to follow a different career path.

When I was hired, I was told I needed five different sport coats; one for each non-weekend day. Since I only owned one suit and was being paid $130 a week, my choices were limited. Woolco had a sale on polyester models for $17 a piece, and I bought five different colors. A lot of polyesters gave up their lives so I could have those sport coats....

I happened to stand took close to a hot light bulb while wearing my dark blue one, and it melted a big spot on the side. Fortunately, you couldn't tell it on air, so I wore it for years. Once I shunned the necktie and wore a turtleneck under my sport coat. No one told me I couldn't, and I continued to wear turtlenecks for several weeks. Then one Sunday evening, Merrell Stebbins, our weekend anchor, did the news wearing a *leisure suit*. A directive from the station manager went out the next day, and I was back to wearing ties.

One of my sport coats had a wild hound's-tooth pattern, and it really needed to be worn (if at all) with solid color pants. One day after finishing my early shift, I headed home and grabbed George's blue sport coat by accident. It shouldn't have been a problem because we wore the same size, but George was wearing his loudest checkerboard pants that day. These babies would not have been out of place on a golf course.

Anyhow, the combination of these two items of apparel was painful to see. George didn't discover what I had done until he was about to go on air. He pleaded with Jerry, our director, to be sure he instructed the cameraman not to show the pants. When newscaster Bill Henry turned to George, he couldn't resist. "Always the trendsetter, George Wooten is breaking new fashion ground this afternoon". The camera pulled back and panned up and down as George just stood there with a frozen grin and clenched teeth. The guys living on park benches downtown were more fashion coordinated than George. He sure caught the viewers' attention that day. He prided himself on his appearance, and it took some big apologies on my part to mollify him.

One Christmas, Connie gave me a new yellow necktie. Since my boss took the day off, I filled in on the 6:00 pm news. The anchor wrapped up his first segment of news and turned to me, saying, "Now turning to the weather with Pat Mo... WOW! Some of you got new color TVs for Christmas. Here's your chance to fine-tune the color. Set your TVs for the deepest yellow you can, and you'll be right on the mark". Connie was watching the show from home and wasn't amused. I never wore that tie on the air again.

I mentioned earlier that I had the opportunity to meet some pretty well-known folks. Robert Conrad of Wild Wild West and Black Sheep Squadron was the nicest of the bunch, but Hogan's Heroes' Bob Crane was clearly the funniest. NBC was promoting a new show where he played an insurance executive who made a midlife change and went to medical school. In the episode that had aired the night before, his patient was the actor John Astin who played a gay architect. This was groundbreaking in the mid-70s. The episode was every bit as funny as the stuff Will and Grace came up with a few decades later.

Before the mid-day news, sportscaster Bob South and I were chatting with Crane in the lobby and discussing his show. He talked about how much the producers wanted to do with the gay theme, and he was pretty satisfied with the way it came out. A short time later, Bob and I were sitting next to each other at the large news desk, getting ready for our show.

Please understand this was not a "happy news" day. Bob's main story was about a long-time local golf tournament being cancelled for lack of funds, and my lead story was about a continued drought. If you were caught watering your lawn, you could be fined. We had certainly had more cheerful days on the set. I went through my segment without incident, and then, as the camera began panning down the Forecast Board, I sat back down next to Bob. In front of me was a monitor. I used to see the same thing the viewers were watching. Unseen by anyone else in the studio, Bob reached over and began squeezing my hand. I busted out laughing, and he just squeezed harder. I was now laughing hysterically. When the camera finally came back on me, I just said, "Sorry folks, I just couldn't help it". Bob sat there stoically and simply stared at me for a few seconds before beginning his segment. As soon as we got off the air, I punched him hard on the shoulder and vowed to get even but was never successful.

Pat Moore
Morning Meteorologist
WFLA TV NBC Channel 8
Tampa 1973-76

In all, I spent 3 ½ years on the air. In 1976, Connie announced that her job had been transferred to Washington, DC, and I was welcome to join her and our two-year-old daughter there. Two years later, she accepted another transfer to Hartford, CT and I made that move as well.

Remedial Dating

I've been married for 53 years, although honesty compels me to admit that I had eight years off for good behavior. In 1993 my marriage of a quarter century came to an abrupt halt. Those of you who have experienced a dissolution of marriage know that the term "messy divorce" is redundant. We started out to resolve things amicably but made the mistake of hiring barracudas to represent each of us. Despite their worst efforts, we were able to stay in touch on civil terms in later years.

Being single again after so many years presented a dilemma. The skill of riding a bike is something you will never forget. You can go decades without straddling one, and you know it will come back to you in an instant. Dating is not biking. Whatever skills one might have acquired over time are conspicuously absent when thrust back into the dating world. I'm a perfect example. I couldn't grasp the idea that I was no longer married and even continued to wear my ring for close to a year. After fifteen months, my oldest daughter, who was a student at Tufts University, asked me if I had begun dating, and I replied, "No, no one has asked me". She promptly arranged for me to meet her roommate's mother. It was a nice evening out, but other than chatting about our kids, conversation tended to lag. Clearly, I need a refresher course. The last time I had been on a date was during LBJ's administration.

I went to the library to see if they had any instructional manuals and found one that had lain undisturbed so long it was covered in dust. The first two chapters were titled "Necking" and "Petting". Computer dating had been in the news, and I decided to give it a try. My Dad had given me a cast-off Radio Shack TRS80 DOS computer with a 1200 baud dial-up modem. Prodigy was the service of choice at the time, and I began to peruse the "bulletin boards". The internet as we know it may have existed but certainly not to the extent that it is today. By logging time on the Connecticut-based boards, I "met" some women who seemed interesting, and I began asking them out.

The first one lived a little south of Hartford, and we agreed to meet at a bar in a neutral location. Susan was attractive but reeked of nicotine. I've always found kissing a smoker about as enjoyable as licking an ashtray. By way of introduction, she proceeded to tell me about a relationship that had just ended and the fact that her oldest daughter had become a mother and she hadn't seen the grandchild because she and the daughter weren't talking. She then mentioned that the second of three ex-husbands was featured in the current issue of "Tattoo Monthly". I sensed she had a little difficulty with relationships.

The evening was still early, and she suggested we visit a Karaoke Bar. I was game as long as I wasn't asked to sing. I'm so bereft of talent that five of the six New England states have issued injunctions preventing me from singing and New Hampshire currently has the measure before its legislature. It turns out that my date had a pretty decent voice, and I tried not to be offended when she sang "You're So Vain". Susan was a "Para-Mensan" which she explained was one level below the intelligence level needed to join the Mensa society. I was bright enough to know we weren't a match.

Back to the computer. I upgraded to a windows-based PC and returned to Prodigy to discover that chat rooms had been invented. For the first time, online surfers could pick a "screen name". Since I was the oldest active participant in a bulletin board devoted to "Forty Plus Singles", I adopted the screen name of "oldestmember". Shortly after that, I got my first instant message from a woman in Rhode Island who typed, "I have one question. Just how old *is* your member?" About a dozen of us from that Prodigy bulletin board made plans to meet in the Poconos in the summer of 1994.

It was like meeting the e-version of a pen pal for the first time. Over drinks, I proposed a toast to the better half of a dozen failed marriages when someone pointed out that the 12 of us actually represented 20 failed marriages! Ironically, the men in attendance had each experienced one divorce, but most of the women had tallied 2 or 3. I remember wondering why someone would keep putting themselves through so much grief. I recalled a line a co-worker had used. When he was invited to a buddy's fourth wedding, he advised the guy, "Instead of getting married, why don't you just compile a list of people you don't like and buy them houses?"

Back to Prodigy. I struck up an acquaintance with a chemical engineer from Groton, CT. We never actually got together face-to-face, but the chemistry was certainly there.

In the fall of 1994, I met a 34-year-old from Stamford, CT, through Prodigy and we began dating. She was a wine expert and served a nice merlot the first time I was invited to dinner. I love wine now, but at the time, I was exclusively a beer drinker. I told her I liked the wine, and it was the first time I had had a merlot (I pronounced it with a hard T). Despite my gaffe, she made it a point to teach me a little about wines. She told me early on that she had a hard time making relationships last longer than 8 weeks. Six weeks later, she announced that she felt it was time to move on when I blurted out, "You can't! I have two more weeks!"

America Online was gaining popularity about this time, and they had the equivalent of a dating service. You could post profiles and photos, and I took the plunge. I got a response from a woman from my own hometown. As we were instant messaging, we discovered that her son and my daughter had played golf on the same high school team. Small world.

After a fairly lengthy relationship ended, I found myself back at square one. I was nearing age 50 and seemed destined to spend the rest of my life alone. That's when I started going to the Gallery on Sunday nights. Based in Glastonbury, CT, the Gallery is one of those large edifices that can be rented out for birthdays, weddings, and private parties. It was once owned by hockey legend Gordie Howe. Every Sunday, the doors opened to Connecticut's singles. Regulars would even come from Massachusetts and New York. For less than $10, you gained admittance and were treated to a pretty good buffet. A separate downstairs area with a small combo was intended for the older attendees, but there was a fair amount of traffic back and forth between the two areas. Upstairs was livelier with a DJ and a fair-sized dance floor.

You could spend an entire evening just people-watching. After a few weeks, it became easy to pick out the regulars. It took me a while to realize that a significant number of them *weren't* there to meet a soul mate. Sundays at the Gallery were the high point of their social lives. On the other hand, the vast majority *were* on the prowl. And some of them were scary to watch. Over a period of several nights, I observed a young man of Indian or Pakistani background making slow clockwise circuits of the room. Traveling equally slowly in a counter-clockwise pattern was a balding middle-aged man who always wore black leather pants and a long black leather coat. He had a kind of shuffling gait and seemed to focus his eyes on something a few light years away. The two circuit walkers would pass each other twice on each lap, and I kept watching to see if they would acknowledge each other, but they seemed oblivious of each other's wanderings. It finally occurred to me that I should probably quit staring at these two guys if I was going to have any luck meeting women.

One octogenarian who apparently had too much to drink had wandered up from downstairs and asked me if I wanted to dance. I mumbled something about suffering from the effects of "osmosis" and she said, "Poor dear" as she moved on to the next guy. During the course of many visits, I did meet some nice women and struck up some friendships but never asked anyone out for a date. The scary thing was that *I* was becoming a regular. If the following Monday was a holiday, the place was even more packed than usual. On those nights, I'd estimate a crowd of 700. I recall once trying to calculate in my head what that must have represented in fees to divorce lawyers.

Back to AOL. I had been reading the singles ads and found them pretty boring. A few stood out. One woman had created a cryptogram that said, when unscrambled, "If you're smart enough to decipher this, I want to meet you". Another started off with, "Let's get one thing straight. I *HATE* long walks on beaches!" I decided I needed to come up with a creative one of my own.

Loquacious...

... pentagenarian in search of vivacious, veracious, and sagacious lady. Wishing to avoid mordacious, mendacious, rapacious, predacious, and ungracious. Me? Tenacious, and sometimes contumacious. Occasionally perspicacious, often pertinacious, but rarely pugnacious.

Our getting together might just be efficacious......

By the way, it's okay if you're a little salacious. Just call me audacious......

That one brought about forty responses ranging from "Huh?" to "I understood it all without resorting to a dictionary!" The latter lived in Washington, DC, which dating shorthand deemed to be GUD (Geographically UnDesireable).

I was living in an old Catholic church that had been converted to apartments and was upstairs one night chatting with one of the Forty Plus Singles members from Indiana. Suddenly there was a boom from next door. I typed, "I heard an explosion. I smell smoke!" I ran downstairs, and my kitchen was filling up with smoke. I tried to call 911, but the computer was on dial-up, and all I got was a modem sound. I started to run back upstairs when common sense took over, and I ran outside to borrow a neighbor's phone. To my horror, my next-door neighbor's apartment was engulfed in flames. The fire trucks arrived shortly after that, but it was too late. She and her three-year-old son tragically perished.

Pretty much everything I owned was destroyed. It was the 29th of December, and the temperature was 9 degrees. All I possessed was a bathrobe and slippers. For some reason, I had forgotten to lock my car and found a ratty V-neck sweater, some stained pants, and rubber boots in the trunk. Better than nothing. My car keys had burned up, along with my driver's license and all other IDs. My ex and my youngest daughter were firefighters, and after their work was done, I stayed with them. The next morning, I figured the best way to get my life back in order was to go to my office and start making calls. I got a ride to work looking like a homeless person (which I was!) and stepped into the elevator. Our Chief Financial Officer stepped in too and stared at me for a few seconds before saying, "Pat, you're really taking Casual Friday seriously". I didn't explain. A short time later, after he'd heard what had happened, he stopped by my office to apologize, and I told him it wasn't necessary. I needed a chuckle about that time.

I was able to rent an apartment but having no ID or credit cards is a real challenge when trying to buy clothes and furniture. Fortunately, I had kept a checkbook in my office, and it had my name and address inscribed on each check. I cut out an article about the fire from the Hartford Courant, and with that as evidence of my identity, a furniture dealer let me buy one twin bed. Sears agreed to let me buy one suit, one tie, one shirt, one pair of shoes, etc. Eventually, I got my identity back and started my life over.

Back to dating. In all, I had 17 different internet dates. It was pretty evident on a few of them that a follow-up wasn't in the cards, but I don't think I could characterize any of them as a "bad date". I think we genuinely enjoyed ourselves and, in a few cases, did get together for a second date. But nothing really clicked, so it was back to the Gallery. I had struck up a couple of friendships with women who belonged to the same ski club, Mountain Laurel Skiers, out of New Britain, CT. In November of 1996, one of them told me about an upcoming Open House. She said the club required members to pass a ski test (a requirement which has since been dropped). I had learned to ski in Bavaria when my Dad was stationed in Germany in the mid-fifties but, other than one or two times hadn't skied in 34 years. My friend told me this was the premier social club in the state, and that was all I needed to hear.

I attended the Open House and joined the club that night. I bought used skis, boots, and poles and rediscovered the joys of skiing. Better yet, I soon met my future wife. A few of us were staying at the club's spacious lodge in Vermont and decided to sample the nightlife in Killington. Penny Trick was the only woman in the car, and on the way, we discovered we shared an interest in Presidential trivia (one of her four degrees is in Political Science). I then asked her to answer a geography question I had included on a test when I taught the subject at the University of South Florida. If you draw a line due north from the eastern border of the Texas Panhandle, name the states you encounter before you hit Canada. She correctly identified them, and I was suitably impressed.

Shortly after that, I began dating another woman in the club, and that relationship lasted ten weeks. When it ended, I went back to the internet and discovered that a singles function was going to be held at the Charthouse Restaurant in the nearby town of Simsbury. The date was June 6, 1997. I met a woman and was chatting with her when Penny walked in. When she saw me, she said, "What the hell are you doing here"? My conversational companion replied sarcastically, "Nice greeting, Penny". They had known each other for some time. In any event, I explained that I was once again unattached, and Penny and I headed to a bar for an informal date. That was more than twenty years ago. We were married on June 24, 2001.

To be honest, I was in no great hurry to marry again. Ending marriages is so stressful that I think the authorities should make the process of getting married as difficult as ending it. Fewer people might leap into ill-advised unions. Penny, on the other hand, had never been married and was more inclined to change that. She didn't bring up the subject more than three or four times a day, but I sensed it was what she wanted. On a bus trip to New York City late in 1999, she pointed out a DeBeers billboard sign depicting a diamond with the slogan, "What are you waiting for? The *third* millennium?" I got the hint and proposed the following February. I suggested that we fly to Vegas and get married by an Elvis impersonator. Since she had been planning her wedding for 48 years, she immediately rejected that idea, but she surprised me by reaching out to our very talented friend Neal Fisher who showed up at our reception as Elvis. He even sang "Love Me Tender" to her! It was the party of the decade.

Everything has worked out great, and she likes to introduce me to folks by saying, "I'd like you to meet my first husband".

Education

Growing up in a military family meant moving a LOT! Grades 1-3 were in Virginia, 4th grade was in Alabama, and part of 5th in New York. The rest of the 5th plus 6th and 7th were in Germany. Grades 8-9 were in eastern Massachusetts, 10th in Alabama, 11th in western Massachusetts, and finally 12th in Maryland. In my Senior year, I had to take a German class by correspondence through the University of Kansas because my school didn't offer it.

My sophomore year was spent at a school that had the worst sports team name in history. Sidney Lanier High School – the Home of the Fighting Poets! Ironic when you consider that Bart Starr graduated from there.

I'm not blaming my poor academic performance on the frequent moves, but it didn't help. I'm fond of telling folks I graduated in the "Upper 90% of my class". Later in life, my oldest daughter was nominated for a Fulbright. I was in the running for a Half Bright….

I'm not saying my IQ hovered around room temperature, but it paled in the presence of both my daughters. My self-esteem did get a slight boost when I had the top score on standardized tests at USAF Basic Training, and then I graduated first in my class at Weather Observer School. When I scored 90 percentile on the Law School Admission Test, I had to scratch my head and figure out why I was such a poor student. It occurred to me that I had to be genuinely interested in a subject in order to do well in a class. I really enjoyed physical geography and, as a result, maintaining a 3.9 GPA in graduate school.

I mentioned that my daughters were superb students and excelled in academic pursuits. Combined with Penny and myself, we have eleven degrees! Penny alone has four, including a Law Degree.

Early on, I taught the girls the memory system developed by Harry Lorayne and Jerry Lucas in The Memory Book. Lorayne appeared on many TV shows over the years, meeting the audience members as they entered the studio. Live on the air he would ask everyone to stand and then sit when he called their names. He would do this successfully with up to a thousand people. His system employs mnemonics and peg words.

When my youngest daughter was about six years old, we were enjoying a soft drink following a round of golf when I suggested a test to two other guys in the room. We numbered a sheet of paper 1 – 100 and had the guys assign a random word to all the numbers. Their choices ranged widely from a golf tee to Abraham Lincoln. When the list was complete, they asked if I was going to memorize it, and I said, "No, she already has". My daughter recited all 100 items without a mistake. The guys would then ask questions like, "What's item 22" or "What number is the pickup truck?" She got everyone right.

A guest was at the house once, and I asked him to select forty random numbers and write them down. One daughter then recited them correctly, and the other one did the same thing in reverse order.

I still use the system to this day.

Jobs

Most folks choose a career and follow it through to retirement. That seemed boring to me. Following my four-year stint on active duty with the Air Force, I wrapped up my Bachelors and Masters degrees and landed two jobs at the same time: the morning meteorologist at Tampa's NBC TV station and adjunct instructor or physical Geography at the University of South Florida. On most days, I'd do a show, teach a class, do a show, etc.

Following the move to the Washington, DC area, I sold life insurance and worked for an employment agency while attending Law School at night. Another move ensued, this time to Hartford, CT, and I joined another employment agency, becoming a partner.

I opened a branch in Springfield, MA, and ran it for eight years. Over that time period, I amassed a large collection of "classic" resumes. I was astonished at what some people would list on a summary of their employment.

- A professional Proofreader misspelled the word Proofreader on his resume!
- In "Other activities," one applicant stated that he loved to "scream loudly at Stravinsky Ballets".
- Another stated that he was 6' 8" tall and required a place of employment with tall doorways.
- Another "Mojored" in English.
- Typos were always entertaining. Employment at the most recent job was from "1881 to the present".
- Back then, we still asked for dates of birth, weight, etc. One woman put down, DOB July 9, 1954, Weight *8 lb, 11 Oz*.
- One applicant came in for an interview in a new suit. I noticed that it still had the large tag sewn on the sleeve. Apparently, he bought the suit just for the interview and intended to return it afterward.
- There were some really interesting names too. If your last name is spelled Anis, why on earth would you name your kid Harry???
- Some other notable names I remember: Pinky Chuang, Zeditha Cabbagestalk, and Tatsuko Goldfarb.
- In the course of one month, I received resumes from women named Alpha and Omega!

That led to a spot as Director of Sales Recruitment at Advest, a regional brokerage firm. The firm was sold three times in rapid succession, and ultimately everyone but the brokers experienced a layoff. At this point, I was 60 years old and was experiencing age discrimination for the first time. Twenty-seven interviews ensued, but when they realized how old I was, they lost interest. One interviewer asked me my biggest weakness, and I replied, "dark chocolate," just to see if she was paying attention.

The owner of a ski and bike shop in northern Connecticut asked me to design a webstore for him. I did so and ran his online store for the next eight years before retiring. I had been working weekends as the coordinator and pacesetter of NASTAR (the world's largest recreational ski and snowboard racing program) at Okemo Mountain in Vermont during the same period of time. Penny and I retired in 2014 and moved to Vermont to live in a condo we had bought in 2000. I worked summers at the golf course, and Penny taught Nordic skiing in the winter. We both taught courses in iPhone Photography. She also worked as a substitute teacher. These days I sell my photography, pencil drawings, and acrylic paintings. I'm still not sure what I want to be when I grow up.

Cars

In April 1967, the Air Force assigned me to MacDill AFB in Tampa as a Weather Observer. Base Housing was filled, so I was given a room at the Bayshore Royal Hotel. Shuttle service was sporadic, so I found myself needing a car. Base pay back then was $98 a month, so my options were limited. A Major in our Squadron was selling a 1961 Rambler Wagon for $250. He tried to talk me out of buying it, but nothing else was in my price range. I soon discovered his reluctance. I'd pull into a filling station and ask the attendant to "Fill the crankcase and check the gas". That thing went through oil at an astonishing rate. The Major felt guilty and gave me $25 bucks back. I found an amateur yard mechanic who said he'd do a valve job for $100. It took him six weeks to do the job, and then the clutch started to go.

Enough was enough. I limped into a used car lot and swapped it for a '66 VW Squareback with a moon roof which I discovered didn't close fully. The first night I owned it, I parked it on a slight slant. I didn't notice the next morning that the headliner was sagging down, filled with rainwater. The first time I braked, about two gallons hit me in the back of the neck! The electrical system went next, and the local VW mechanic named Klaus Porsch told me, "Herr Moore, you haf vat ve call a LEMON".

I bit the bullet and bought my first new car, a black jade 1970 Ford Maverick (no air conditioning in Florida!).

I've lost track of how many cars I've owned since then, and none of them were particularly memorable, but we did make an impulse buy during the COVID-19 Pandemic. We proudly own a 1999 Corvette Coupe and are using it to visit all 251 towns in Vermont with it. As of this writing, we have 70 to go.

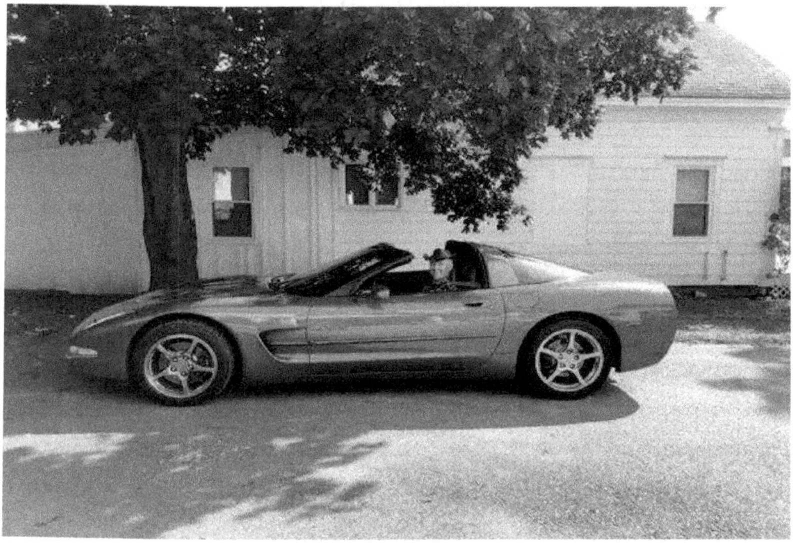

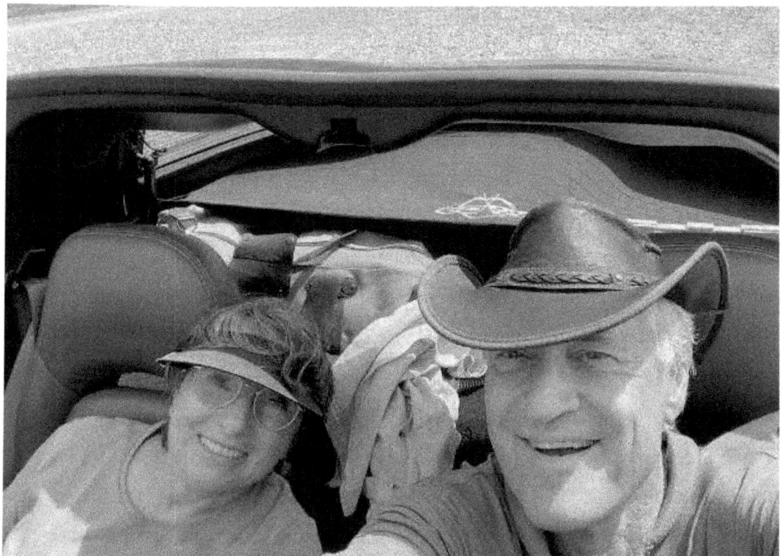

Bodily Harm

Quoting David Bowie, "Take your protein pills and put your helmet on".

Whether we were snow skiing or water skiing, Dad was fond of saying, "If you're not falling, you're not trying hard enough". Well, I try hard! My body has endured more than its share of abuse through the years. I guess I've always pushed the envelope when engaged in sports, and it's caught up with me more than once. As early as four years old, I was trying a balancing act on a fence and fell, landing on a broken bottle. Mom freaked out to see blood gushing from my wrist, but the emergency room folks got it under control, and I got introduced to stitches. I still have a four-inch scar on my wrist.

As I got older, I got interested in gymnastics and had pretty good success in competition before I became a varsity diver for the University of Florida. I have poor vision, but it generally didn't present a problem at outdoor pools. Competing inside was more of a challenge to my vision. In 1965 the Southern Intercollegiate Championships were held at the University of Georgia or Georgia Tech, I can't recall which. The vast indoor pool had been used for training WWII pilots in case they were shot down in water. The building was a gloomy place, with the brightest light coming from the giant frosted glass windows on the wall opposite the diving boards. Competition on the 1-meter board went smoothly, and I took fifth place in the competition. The 3-meter board was another story.

The building's lights, such as there were, were suspended from the high ceiling on long cables and had deflectors so that all the light showed down. When a diver gained the maximum height after leaving the board, he would find himself in the darkness above the lights. A fraction of a second later, he would be illuminated by the lights and the frosted windows. This wouldn't be a problem for someone with good vision executing a simple forward dive, but when Mr. Magoo tries a forward 3 ½ somersault, spatial orientation is conspicuous by its absence.

When it was my turn to dive, I began my paces and hurdle step. The board compressed downward and then launched me upward into a world of darkness. I began my forward spins and prepared to break out of my tuck for a flawless entry. Back home at Gainesville, I always broke for the brightest spot, which in my blurred vision, was the pool itself. On this particular occasion, the brightest spot in the building was the frosted glass window leading to the parking lot outside. Instinctively I straightened out and could not have been more perfectly parallel with the water's surface when I impacted it. If the Olympics awarded medals for belly flops, I would have easily topped the podium.

The impact of slamming a horizontal body onto a horizontal surface from a great height after spinning like a top is not an experience I'd wish on anyone. The shock to the system is indescribable, and the whole body goes numb. I recall a few other divers helping me out of the water and asking if I was all right. Stupidly, I said yes. Saying no would have been the smart move because I still had to do an inward 2 ½. If you know what is commonly called a cutaway, you'll have an idea of what I had to do. Facing the board, I had to jump up and away from the board before beginning 2 ½ forward revolutions at high speed. This time, when I broke for the brightest spot, I smacked the water flat on my back. The pain was excruciating. The same guys as before jumped in and hauled me out. This time they brought in a doctor. My back had acquired a red and white splotchy pattern that made me look like a sunburned Guernsey. I was "grounded" from further competition.

Following college, I ceased diving but kept proficient at gymnastics. I rarely had to buy a beer as I would bet other bargoers I could do a standing back flip. When the beer was done, I'd bet them I could do a standing front flip (much harder), and if I was still thirsty, I'd bet I could do a "which foot" back flip. In the middle of the flip, someone would shout out "left" or "right" and I'd land on that one foot.

I could get enough height on a flip to do it in a pike position and even sort of a layout back flip I called a whip back. That won more than a few beers.

Returning from the 1964 Worlds Fair in NY, I once did a back flip in a lurching club car on a train that had a low ceiling and a narrow aisle. I stuck the landing, but I'm told my feet barely cleared the ceiling.

I didn't always exercise the best judgment as to when to do the flips. Driving home from a bar in Zephyrhills, FL, one night, I got pulled over by a Florida State Trooper. After checking my license and registration, he instructed me to stand on the solid white line and walk forward, placing my feet on the line. After a few steps, I stopped, did a backflip, and landed precisely on the white line with one foot in front of the other. He gave me an incredulous look. After a few seconds, he just shook his head and said, "Get outta here". I was happy to comply.

Pursuing gymnastics resulted in pretty good flexibility even in my fifties, and I sometimes used it on the dance floor.

When dancing to fast music, I develop a case of "Happy Feet". They have a mind of their own, and I just try to keep up. Many years ago, that would sometimes involve doing roundoff, back handspring, and layout backflip before segueing into the Watusi, the Twist, the Locomotion, etc.

Inevitably my pulse would go way up to around 190 bpm and stay there for a disturbing length of time. I really have poor endurance. I once had a stress test, and the cardiologist told me I was a "genetic mutant". He said everything else worked fine, but that exercise would trigger a rapid pulse firsor the rest of my life. Oh well.

In 2022 I visited the VA Clinic in Rutland, VT, to be checked again. They fitted me with a heart monitor to be worn for two weeks. During that time, I worked at the NASTAR ski racing venue at Okemo Mountain, where I've been the pacesetter for the past sixteen years. The job involves hauling a lot of rolled-up safety netting and can be exhausting. The test results indicated that my pulse had exceeded 200 beats per minute on more than one occasion.

I then met with a cardiologist and told him my history. I've always been fast for short spurts but lag a lot on longer distances at speed. I have friends who can bike a hundred miles at more than 20 mph, while I'm hard-pressed to maintain 14 mph. The doctor gave me this analogy. "A greyhound can run as fast as 45 mph for lengthy distances. A cheetah can hit 75 mph for a very short period of time. Your friends are greyhounds. You're a cheetah. There's nothing you can do about it."

I didn't go to the gym this morning, and that makes seven years in a row....

Sports

In the mid-seventies, I was contacted by a Doctoral candidate doing research on individuals competing in dangerous sports. He reached out to the top fifty finishers at that year's National Parachuting Championships and asked us if we take a battery of tests. I guess he was trying to find out if we had a death wish! I agreed to participate and was sent the materials, which took a long time to complete. The only one I remember was the Minnesota Multiphasic Personality Inventory. He promised to send me my results, and they arrived more than a year later. The chart compared me with the general population in a wide variety of personality traits. In all but one, I was as close to the average as you could get. The exception was "Desire to Win". I was almost off the charts. I'd never thought about it, but I admit I don't like to lose.

Over the years, I've participated in several sports with varying degrees of success. In the fifth grade, I was ten pounds under the minimum weight for Pop Warner Football, but they let me play anyway as a Free Safety. Picture a 65-pound punt returner. Little League and soccer were next, and then I discovered gymnastics. Being the smallest student in a two-town high school (boy or girl), I was drawn to that sport and pretty much taught myself. By the tenth grade, I could do standing back flips and continued doing them until age 50. As I mentioned earlier, I'd win beers doing flips in bars. I would also win bets by having someone stand on my stomach while suspended between two chairs' backs.

I enjoy competition regardless of the sport. I pursued competitive skydiving in Accuracy and Style (which involves a series of speedy turns and loops in freefall). I competed at the National Championships four times and was the Florida Parachute Council Overall Champion three times.

In 1976, I retired from skydiving and briefly got involved with rock climbing before discovering golf two years later at the age of 42. I had a slow start in that sport but managed to lower my handicap each year. In 1996 and 1997, my handicap dropped to 5.8, and I won back-to-back Club Championships.

Penny and I spent our honeymoon in the UK, and I got to play where the game of golf was invented - the Old Course at St. Andrews on the 4th of July, 2001. What a thrill! It was a windless day, and I managed to miss all 112 of the course's nefarious bunkers. I needed to finish birdie-birdie to break 80 but three-putted both greens.

Golf is still one of my favorite pursuits, but sadly, my game has gone south. Despite holing out a 2 iron from 200 yards and a half dozen other holeouts from the fairway, I've never had a Hole in One. Not long ago, I was playing with my friend Jim Remy who is a Past President of the PGA of America, and when we got to the 17th hole, I said, "This round has been boring. Put it in the cup". Incredibly he did! He calmly announced that it was his seventh career ace, and I told him we now collectively had seven.

* * *

For a while, I was riding a unicycle as much as 12 miles at a stretch. You can't believe how many people think they're the first to cleverly yell out, "Hey, you lost a wheel!" I used to respond by telling them I was buying a bike on the installment plan, but I switched to saying, "It's worse than it looks. It started out as a tricycle!"

Penny and I used to do annual 100-mile rides on bikes at Tiverton, RI. One year she did 50 and finished early. She had a pretty good idea when I'd finish and met me a quarter mile before the finish line with my unicycle. I switched to it and came to the finish while a friend played the theme from Rocky on his boombox. That got everyone's attention! These days I only haul the unicycle out on my birthday.

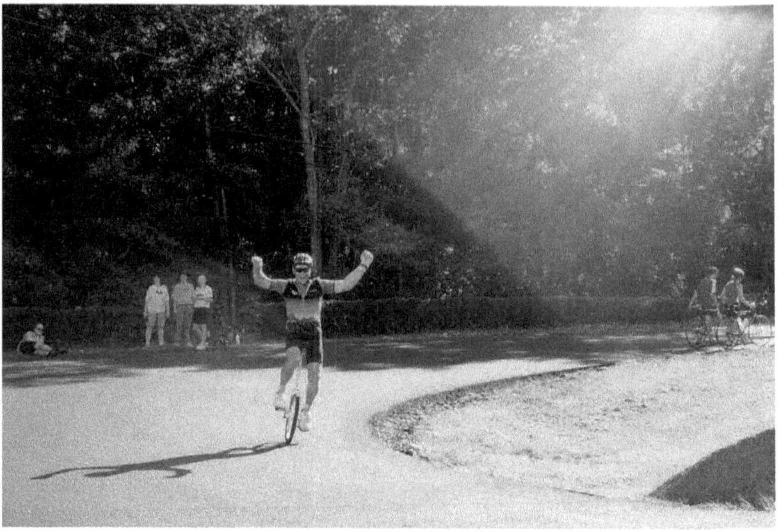

Injuries

In 2000+ parachute jumps, I tore an ACL and had a few sprained ankles but never broke a bone. Then I got into ski racing....

- Straddled a gate: broken ribs that resulted in pneumonia
- Tip of a ski hit a gate: broken foot
- Snowboard crash: broken tibia-fibula
- Didn't see a rope across a trail (shouldn't have been there) and got "clotheslined" in the mouth: Concussion and broken collarbone
- Hit the electric eye at the finish line and broke ribs.
- And unrelated to winter sports, I experienced a UPD (UnPlanned Dismount) from one of my unicycles and broke my hand just before a trip to Hawaii.

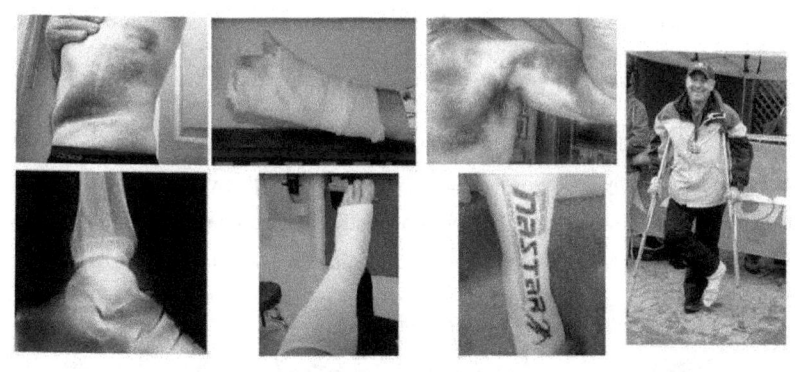

The injuries haven't deterred me. All six of my knee surgeries (including a Total Knee Replacement) took place in the month of May, so the upcoming ski season wouldn't be impacted.

Could age have anything to do with the frequency of injuries? My inner child is still 18, but my birth certificate says 1946. What exactly determines when we are officially old? Is it the onslaught of aches and pains upon arising in the morning? The need to take afternoon naps? Forgetting your best friend's name? I officially became old when I noticed that hair grew better from my ears than the top of my head. I can live with eyebrows that would make Leonid Brezhnev jealous, but fuzzy ears are a sure sign that you're wrapping up the back nine. I'm not ready to throw in the towel. My friend Dick Cole still races on skis and snowboards at the age of 88! I want to be just like him when I grow up.

I had skied as a kid but had stopped in 1962 and didn't resume it until 1996. I took up snowboarding at the same time, and I still split my time between skiing and boarding to this day. After a few years, I stumbled across the NASTAR Race Course at Okemo Mountain in southern Vermont.

NASTAR is the largest public ski and snowboard racing program in the world. I wasn't hooked immediately but caught the bug over time. My first trip to the NASTAR Nationals was to Park City in 2004, and I've attended pretty much everyone since. In 2006 I became the NASTAR coordinator and pacesetter for Okemo Mountain and continued to do that to this day.

Pick any top-level winter sports athlete, and you're pretty much assured to discover that injuries have impacted their careers. Despite the fact that my level of performance is many levels lower, I accept that injuries create interruptions in my racing career. That doesn't quell my desire to continually strive to improve and sometimes win. Once in a while, the stars are aligned, and good things happen.

In 2008 I made a little history by becoming the first individual to be Age-Group National Champion in both Skiing and Snowboarding at the same time. I've won the Snowboard title 14 more times, but I acknowledge that I've probably outlived most of my competition! This article appeared in Ski Racing Magazine. That's downhill racing legend AJ Kitt on the right.

Champions crowned in Steamboat
EVENT ATTRACTS RACING FAMILIES AND WORLD-CLASS PACESETTERS
BY WINA STURGEON

Pat Moore, 62, Newington, CT., made NASTAR history by winning the gold in both skiing and snowboarding. Here he poses with pacesetter AJ Kitt at the Race of Champions finish area.

Ski bindings employ something called a DIN setting which is calculated to allow you to have the boots released when necessary to prevent injuries. They're supposed to stay attached at all other times. When I first bought a pair of racing skis, I filled out the form asking for my age, weight, and skiing ability so the ski tech could adjust the bindings accordingly. I guess I was underestimating my ability because I managed to slide out of the bindings on turns in the first few races. Peter Englert photographed one of these mishaps at the finish line of a race at Okemo Mountain in southern Vermont.

Eventually, I started tightening the DIN settings and kept the skis on.

Given a choice between skiing and snowboarding, I always opt for the latter. Unlike most snowboards, mine are designed for racing. They are longer, narrower, and use plastic boots that resemble ski boots. I used to use a step-in binding that occasionally wouldn't release. Once in Davos, Switzerland, I had to pee really badly, and the rear binding wouldn't release. In a panic, I hit on this solution. I got some funny looks entering the bathroom, but it solved the problem. When we got back to the states, I gave away the bindings.

Pushing the envelope definitely is appealing. When on a ski club trip to Davos, Switzerland, we encountered what looked like a race course with no gates. We discovered that it was a speed test. The skier would exit a Start House and then go as fast as he or she dared. Toward the end of the course was a set of electric eyes ten meters apart. They would be used to calculate your speed. Our club's fastest skier hit 48 mph. I was clocked at 55 mph on my snowboard. Not the smartest thing I've ever done. I was already in my sixties at the time.

Speaking of snowboards, I foolishly tried to negotiate a halfpipe shortly after I had taken up boarding. As I approached the edge of the halfpipe, I ran out of experience and wiped out big time. My head slammed into the icy surface pretty hard. Fortunately, I was wearing a helmet, but I had a hard time remembering names the balance of that day.

The Wolverine Bar

The Wolverine Bar in Zephyrhills, FL, run by two sisters, was a popular hangout for skydivers at the end of a day of jumping. One sister, whom we called "Aunt Margaret" had been an ambulance driver in the First World War. They were pretty tolerant of us.

Wikipedia tells us that "The high point of 'streaking's pop culture significance was in 1974 when thousands of streaks took place around the world." We were ahead of our time…. On October 21st, 1967, I led the World's First Nine-Man Streak through the bar.

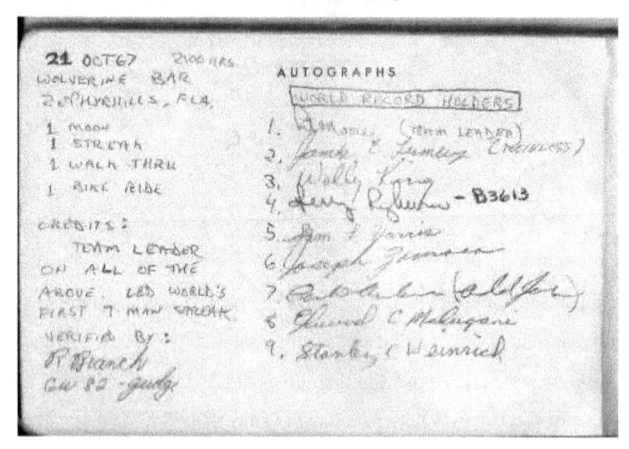

As you can see from the entry in my skydiving logbook above, one person even rode a bike, and one sauntered at a very sedate pace. We were told that an 8mm film exists, but I've never seen it.

Dead Ants was a popular game there. Out of the blue, the designated Antmaster would shout out, "Dead Ants!" and everyone had to hit the floor on their backs with their feet and hands in the air. Last one on the floor had to buy the next round. Just saying the words quietly could trigger the response, so we took to referring to it as "expired pismires". I recall once sitting on a bar stool drinking a beer when the call went out. I put my foot on the bar and shoved back, landing flat on my back. Happily, I was unscathed and avoided springing for another round of suds.

I had been going to the Wolverine before I had turned 21, and Aunt Margaret once caught me drinking beer that I had poured into a Coca-Cola can.

The drinking game of Cardinal Puff attracted a lot of attention. To earn the ranks of Cardinal, Bishop, or Pope, one had to drink beer, recite precise phrases, and perform a series of body movements in a prescribed manner. Making even the slightest mistake earned the command to "drink up" and start over. Becoming a Pope involved an entire pitcher of beer. On August 24, 1967, two months before my 21st birthday, I earned Pope status and subsequently became known as Pope Patrick the Minor. Sadly, the Pope who signed me off, Joe Pelter, was later killed in a DC-3 crash that took the lives of the crew and eleven members of the Golden Knights, the US Army precision skydiving team.

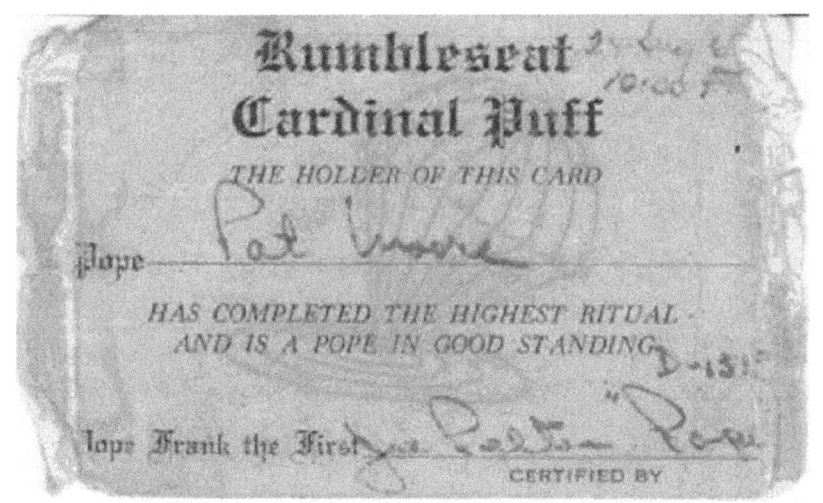

Speaking of streaking, I was competing at the National Collegiate Parachuting Championships in Deland, FL, a few years later when a friend did a solo streak across the large banquet hall only to discover that the exit door was locked! He did a quick about-face and ran into a waitress. In a panic, he ran back to the entrance and escaped. Once he had his clothes back on, he returned to the banquet hall to apologize to the waitress. She said in a loud voice, "It was no BIG deal" to the laughter of everyone there.

Fast forward to the early eighties. I used to listen to Paul Harvey while driving to work. One morning he reported that an unidentified man had streaked the Illinois House of Representatives. When I got to work, I called his show and told his assistant I knew who the man was. When she asked who, I said, "The Streaker of the House!" Paul Harvey used that line in his midday show and credited me with it.

Vehicular mishaps

In sixty years of driving, I've only had one mishap, but it was a doozy. Actually, if you count me backing into the door of a Cadillac a week after I got my license, there were two, but it's the latter incident that bears relating.

When I was working as the morning meteorologist at WFLA-TV, I had to get up at 2:30 each morning to drive from our home in Brandon, FL, to the studio in Tampa. My first show wasn't until 6:30 am, but it took a lot of time to complete a forecast and prepare for the show.

I was driving our 1973 Chevy Van and was approaching a railroad crossing when I was horrified to see an oncoming train's lights to my left! The crossing's warning lights had failed. I instinctively slammed my foot on the gas but realized too late that a car in front of me had seen the danger and stopped for it. My van slammed into it, and every loose item in the rear van, including two parachutes, flew forward, hitting the windshield. The impact had knocked both our vehicles on the tracks with the locked bumpers right in the middle. There's no way the train could have stopped in time. Fortunately, my engine was still running. I jammed the shifter into first gear and burned my tires, trying to push both vehicles forward. It worked as we got clear of the tracks just as the train rumbled by with a warning blast that would wake the dead.

I extricated myself from the van and ran forward. The woman in the car was shaken but happily unharmed.

It took a very long time for the police to arrive. When I had finally filled out the forms and arranged for a wrecker to pick up the van, I was getting dangerously close to missing my show. The cop gave me a ride to the station, where I ran onto the set and connected my microphone with seconds to spare. Fortunately, Paul Catoe, the evening meteorologist, had left everything in place from the night before: maps, forecast board, etc. I ad-libbed my way through the show with no one suspecting I was totally unprepared.

I remember thinking later that someday I need to write a book.

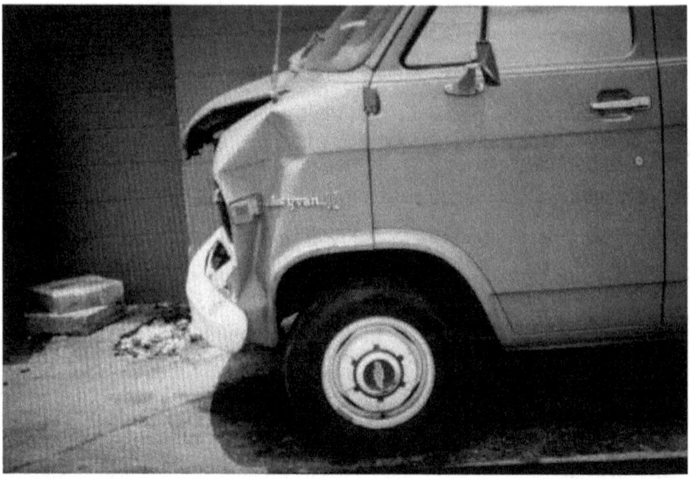

The Blues Brothers

Our ski club is active in all four seasons enjoying golf, biking, hiking, dances, parties, etc. For a number of years, we competed in (and frequently won) a trike race in Killington, VT, dressed as the Blues Brothers and Blues Babes. I adopted the persona of Elwood. Penny and I even attended a performance of "A Tribute to the Blues Brothers" in London in 2001 and were pretty sure it was Dan Aykroyd who briefly walked on stage to wave to the crowd.

Over the years, I've worn that outfit skiing, snowboarding, dancing, rock climbing, surfing, and unicycling. I even climbed the podium at the NASTAR National Championships to get my snowboarding medal dressed as Elwood. I still have the outfit and am waiting for the call, "We're getting the band back together".

I like a challenge. When I worked for the brokerage firm Advest, we occasionally held team-building exercises. At one high ropes course, I asked if I could repeat the challenge blindfolded. I managed to get through the course, but it didn't go over well with my boss, who struggled with the course itself.

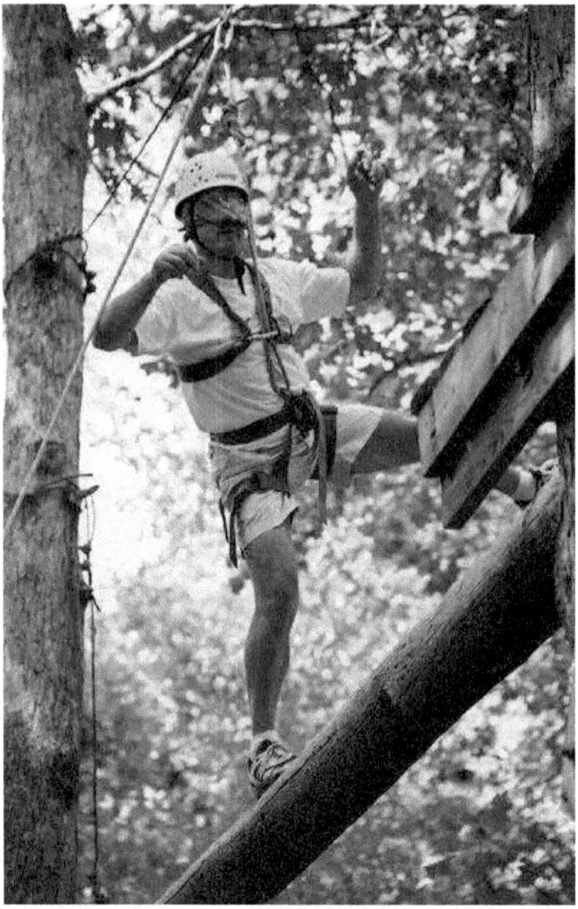

Horse Race Handicapping

I am incredibly inept at picking the ponies. Each year we make a day trip to Saratoga Springs, and on the last few trips, I just used the opportunity to take pictures. When I did place bets, my picks were so bad that they defied the odds!

In 2011 Penny and I, along with friends and my oldest daughter Heather drove to Baltimore to stay with Penny's sister. I volunteered to babysit the grandkids while everyone else headed for the Preakness. Penny was well aware of my incompetence in picking winners but told me to place a bet anyway.

This happened to be the same day, May 21st, that Televangelist Harold Camping predicted that the world would end. He called it Rapture Day. Using that as my inspiration, I scanned the slate of horses scheduled to race. Nothing leaped off the page until I noticed one of the jockeys was named Jesus! Penny said, "it's pronounced HAY soos". I said, "doesn't matter. Ten dollars to win". She then said, "His horse Shackleford is a 13:1 long shot". "Doesn't matter." I repeated.

You guessed it. Shackleford and Jesus thundered home in first place. Penny came home from the race with my winnings: $136 on the 136th running of the Preakness. She dumped the money in my lap and said, "You're SO going to hell".

Stratonauts

In 1960, Joe Kittinger jumped from a balloon at a height of more than 100,000 feet. That record stood until October 14, 2012 when Austrian Felix Baumgartner jumped from 128,000 feet. Kittinger was his Ground Controller. Two years later, retired Google Executive Alan Eustace broke that record with a jump from 135,000 feet. What do these events have to do with me? Stay tuned….

Earlier in 2012 I had written Al Roker and asked him if he'd do a favor for a former NBC weatherman. A week after Felix's record jump, Penny and I were in the crowd at the plaza outside the Today Show and I held up a large sign that said "My Child Bride Just Turned 60!" He came over to wish Penny a Happy Birthday. We have the video. Coincidentally, his guest on the plaza was Felix Baumgartner. We had watched Felix's jump live a week before. I got Felix's attention and we had a nice chat (and got pictures). I later did a pencil drawing of Felix and Joe and sent it to a friend of Joe's who gave it to him. Two years later, retired Google Executive Alan Eustace broke Felix's record and Alan's sister asked me to do a drawing of Alan and Joe. Alan invited Penny and me to attend the premier of a documentary about his jump at the Tribeca Film Festival where we met him and got more pictures. We hope to meet Joe at some point. He wrote me a nice letter thanking me for the pencil drawing and I have an autographed copy of his book "Come Up and Get Me". I am in awe of what these men did and more than a little bit jealous.

The Stratosphere Club

Joe Kittinger **Alan Eustace** **Felix Baumgartner**

Our Travel Channel

Over the past twenty-plus years, Penny and I have enjoyed a lot of traveling. We've been to Europe sixteen times alone. We started recording videos and uploading them to Youtube. Viking Cruise Lines asked if they could feature one we had done of a Rhine Cruise on their Facebook Page, and we agreed. In one day, the video had 11,000 views! Shortly after that, Youtube invited us to create our own travel channel. To date, "Pat and Penny's Travel Channel" has uploaded more than a hundred videos and has more than a thousand subscribers. I invite you to visit the channel at bit.ly/patpennytravel and hopefully subscribe.

We love to travel by air, sea, and land. This map represents just one year.

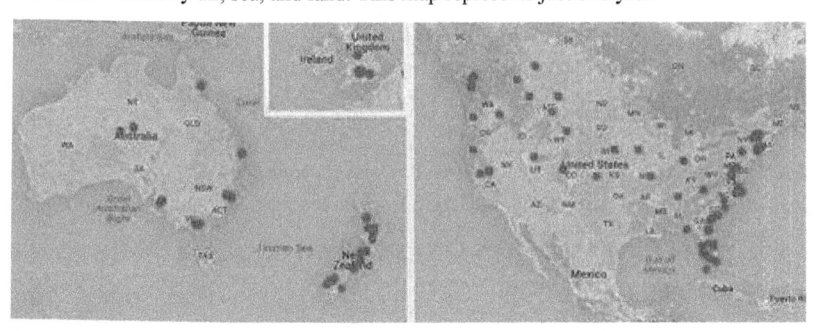

We've done several ocean and river cruises and look forward to many more. We were driving through Florida a few years ago, and Penny said she wanted to visit Key West. If you've never done it, it's a LONG drive. I got on the internet and found a last-minute cruise out of Miami to the Bahamas, Coco Cay, and Key West. We enjoyed five nights of great touring, lodging, and food for $275 apiece. We had a full day in Key West and hit all the major attractions.

Road travel has its own attractions. I've visited every state except South and North Dakota. Penny lacks only the latter.

Here's a snapshot of one year of on-road travel. The jagged line below Florida is the short cruise I mentioned above.

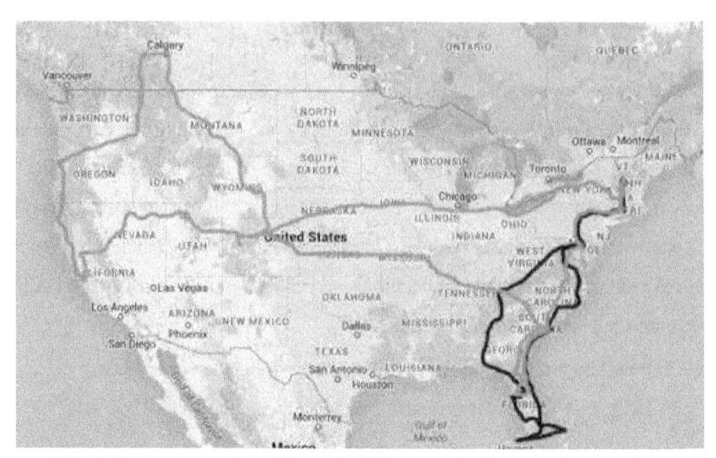

Penny and I had always wanted to take a trip "down under," so we started planning for a visit to Australia and New Zealand. She asked how long a trip I was contemplating, and I said, "Three weeks." She replied, "I was thinking three months"! We "compromised" and did three months in 2015. I'm glad we did. It was a remarkable journey to an extraordinary part of the world. We spent time in Sydney, Melbourne, Adelaide, the Outback, Cairns, Brisbane, Auckland, Queenstown, and Wellington.

A trip of this magnitude requires a lot of planning. To make it affordable, we started saving in a unique way. Every evening when we got home from work, we'd retrieve all the five-dollar bills we had from our billfold and purse and insert them into a slot in a large sealed cardboard box. Penny was fond of saying that if the condo caught on fire, that was the first thing to be tossed out a window. After 4 ½ years, we opened the box and counted ***Eleven Thousand Dollars***!

We used air miles to make the trip (Business Class on one long leg!) and stayed mostly in AirBnBs. The US Dollar was strong against the currencies of both countries. Penny scored big time when she discovered that you could "reposition" vehicles. In New Zealand, many visitors fly into Auckland and rent RVs to travel to the South Island. As a result, rental companies are faced with too much inventory in the south. We were able to rent a campervan in Christchurch for only a dollar a day! They even paid for our ferry crossing. The drive took fifteen hours, and they gave us five days to complete it. We stayed at campsites on the way north and had a wonderful time. If you can handle a five-speed manual transmission while driving on the "wrong" side of the road, it's a phenomenal way to travel inexpensively. We even got to reposition a car from Queenstown to Christchurch.

Some of the friendliest folks on the planet live in Australia and New Zealand. A few months earlier, we had stayed at an Airbnb in London and met a couple from Melbourne. When we mentioned that we would be traveling there, they said, "You'll have to stay with us." They even picked us up at the airport, and we spent some enjoyable time at their hilltop home in the Yarra Valley wine region.

The Outback is a fascinating place. We did a walking tour around Uluru (formerly Ayer's Rock) and slept under the stars. In King's Canyon, we retraced the steps taken by the characters in the movie Priscilla, Queen of the Desert.

We met a lot of German visitors but strangely almost no Americans. We had been there nearly two months before we met a guy from New Jersey at Cairns, close to the Great Barrier Reef. He was my instructor in flyboarding, a fun sport where a jet boat's propulsion provides you with enough lift to fly out of the water. I even got to do some backflips.

Penny and I visited Queenstown on New Zealand's South Island. It calls itself the "Adventure Capital of the World" with justification. On our bus ride to our jet boat ride, we drove over the seventh most dangerous road in the world. Scenes from The Lord of the Rings were shot there. The boat ride was a blast.

We took a day trip to Milford Sound to see the extraordinary waterfalls. From Queenstown, it was a three ½-hour bus trip with the driver lecturing about sheep the whole time. The boat ride was as scenic as promised, but Penny began to feel the effects of motion sickness. On the lengthy bus trip home (this time being lectured about dogs), the queasiness became too much. She "tossed her cookies" in the small lavatory down a half set of stairs, but she didn't feel any better and repeated the trip. Conditions in the lavatory were so bad now she couldn't face it a third time and was desperately looking for a receptacle. Nothing was at hand, and in the nick of time, I flipped my baseball hat upside down. Happily, that was her last bout with nausea.

We had a rest stop a few miles down the road, and I was going to wash out the hat, but she said, "Hell no. We'll get you a new one". She said I definitely got "Hero Points" for my action. It was a black hat with Paratrooper Wings on it. I bought it in Fayetteville, NC, at the Airborne Museum. On our next drive south, we made a detour to Fayetteville and bought a replacement.

***

High on my bucket list was a bungee jump. My brother Mike had done a very high one in Costa Rica, and I really wanted to experience it myself. I remember reading about a guy in Alabama who was going to do one from the roof of a coliseum and didn't calculate the distance properly. He was 28.

In 1989, New Zealander A. J. Hackett created the first commercial bungee (spelled Bungy in NZ) in Queenstown. The country's highest operation at 440' is at nearby Nevis Gorge. We were touring the South Island just after my 69th birthday, and I couldn't wait to try it. I even bought a GoPro just to take a selfie video during the jump.

The launch point was a fixed platform positioned over the gorge. When our names were called, we hiked to a small cable car and shuttled out. It reminded me of watching the crew of the Challenger taking their last walk to the Shuttle launch pad. The other jumpers were young, and I could sense a mixture of excitement and dread in them. Over the years, I had jumpmastered hundreds of first jump students from airplanes and witnessed a wide range of behaviors. The three going ahead of me exhibited pretty much the same degree of fear. Frankly, they were terrified.

When the staff connects you to the harness, the primary anchor point is around your ankles, and they lock together very closely. Picture a prisoner shuffling with shackled ankles. Each jumper had to wiggle-walk in tiny steps to reach the end of the platform. The scene conjured up images of pirates walking the plank three centuries ago. At the "GO" signal, all three of my fellow students stood motionless, bolstering their courage. Eventually, they kind of collapsed off the step flailing with their arms.

Remember, I had logged 2000+ parachute jumps when I was younger. I didn't know if I'd feel any trepidation stepping into nothingness. When I got to the end of the step, I was surprised to find that I wasn't fazed. Way below me ran the Nevis River with steep cliff walls nearly as high as the launch point. Maybe it was the act of making sure the GoPro was running, but I couldn't wait to go. The instructor made sure I looked to the right to smile at their video camera. I grinned at it and then asked if I could go. He said yes, and I launched myself out in the best swan dive I'd done since I was a varsity diver at the University of Florida. Accelerating at 32 feet per second per second, I turned my head to the left and grinned at the GoPro. In the video, you can hear me laughing. The rope stretched smoothly, and my downward plunge ended just above the river. That was followed by a series of up-and-down bounces between the cliff walls before dangling upside down.

Once you've stopped bouncing, you're supposed to tug on a yellow lanyard to switch out of a head-down position. No matter how hard I tugged, mine wouldn't release, so they had to haul me back to the launch point upside down (see middle photo below). I was laughing so hard I didn't mind.

That was one more item checked off my bucket list. Give some thought to trying bungee jumping. It's definitely a blast!

South Africa

In 2019 we traveled with friends to South Africa for safaris in Zimbabwe and Botswana. We experienced "Glamping" (Glamour Camping) for the first time. The walls may have been made of canvas, but the accommodations approached five stars. In all, we stayed at three campgrounds and saw an abundance of elephants, hippos, zebras, crocodiles, and giraffes. We had a few lion sightings and a close encounter with a leopard. Our guides had an uncanny ability to find these animals, and we highly recommend trying a safari at least once in your life.

I managed to leave a GoPro 5 in Botswana. The campsite employees found it. They said it would cost $200 to mail it to the states. I looked up the rates, and they were right. I had won the GoPro in a video contest, so I told them to keep it and bought a newer model.

Dubai

On our way home from South Africa, we did a three-day stopover in Dubai. From the tallest building in the world to the largest mall, everything is the biggest and the best. We started our adventure, thereby enjoying an indoor ski slope. The outside temperature was 110 degrees, but it was 24 degrees inside. Getting to the top was accomplished with a modern four-passenger chairlift or a T-bar. The snow was surprisingly good. Penny hadn't skied downhill in twenty years (she teaches cross-country skiing) but did very well. I rented a snowboard and spent time carving or riding in the terrain park. The novelty wore off after a couple of hours, but it was well worth the time and effort.

* * *

Dubai's ten-million-liter aquarium in the mall is home to 400 sharks. I signed up to Scuba Dive in the tank. Since I'm not dive-certified, I was given a two-hour class by a guy with a thick Kuwaiti accent. I'm hard of hearing, but I got most of what he was saying. He did tell me not to hold my breath and not to let bubbles hit the undersides of sharks. Soon after I got in the tank, I was directly under a huge shark and needed to either hold my breath or blast him with bubbles. I opted for the latter, and he just swam away.

We signed up for a trip to the desert, complete with high-speed rides up and down the dunes. Our Pakistani driver picked us up at the hotel and excitedly announced he had passed his training the day before. We discovered he had been using that line for seventeen years. We also rented ATVs, rode camels, and attended a nighttime show. I even rode a snowboard down the dunes.

Back in Dubai, we ascended the Burj Khalifa, the tallest building in the world, and donned Virtual Reality glasses for a virtual base jump. When I used to do accuracy parachute jumps, I'd exit the plane from 2500 feet. This building is 2727 feet tall!! At the base of the building, there is a nightly fountain show that rivals anything Las Vegas has.

We took a harbor tour in a boat that, thankfully, was air-conditioned. The captain pointed out an apartment at the top of a building that rents for $30,000 a night! The harbor has a Ferris Wheel that is taller than the London Eye!

On the second day, I spotted a westerner in his fifties who seemed out of place. Then it occurred to me that (present company excepted) everyone in sight was young! Expatriates and immigrants make up 88.5% of the population of Dubai, and they are young. The remaining 11.5% are Emiratis and have enough sense not to go outside in blistering heat. That's what servants are for. So for a three-day stretch, I was easily the oldest person seen in Dubai.

Airbags

We live very close to the Okemo Mountain ski area. In the summer, their Adventure Zone has lots of attractions, including a kind of roller coaster, a high ropes course, and a zip line. They used to have a giant airbag and a thirty-foot high platform for jumping. You're supposed to jump off the top, assume a sitting position, and land on your butt. That seemed too boring for me, so I dove off and did half a barrel roll before landing on my back. I figured they'd kick me out, but they were okay with it, and I did it a couple more times.

They used to have the airbag at the bottom of one of the ski slopes. Skiers would launch off a ramp built up with snow and do a flip before landing on the bag. I was going to try a backflip on one of my snowboards, but the airbag didn't return the next winter season.

Drone

To alleviate boredom during the COVID-19 lockdown, I bought a drone. Regulations require that you have to register any drone weighing 250 grams or more, but I found one weighing 249. The downside was that it didn't have a collision avoidance feature. The first day I had it, I managed to fly it into a tree, but it emerged unscathed.

I got better at flying it and logged quite a few hours. We did a day trip to southern Vermont to visit a large waterfall and get some drone footage of it. I didn't pay attention to the battery level and got a "Low Battery – Land Immediately" message when it was still over the waterfall. Not only that, I had to essentially "thread a needle," flying it through a small gap in trees. If I had been thinking, I would have aimed the drone away from me, flying it backward so that moving the control stick right would cause it to fly right. Nope, all the controls were reversed as I brought it quickly back, envisioning my expensive toy being lost forever. I got it as far as the trees with the battery almost exhausted when, you guessed it, I hit the wrong control and flew it into a tree right on the edge of the waterfall. It hit the ground with the props still spinning. I ran around the tree and climbed down to the edge of the waterfall. One foot slipped, but I caught myself in time and retrieved the drone. The video ran the whole time and caught my near fall. Penny was sitting on the edge of the waterfall, watching all of this. There was a long steep trail to the bottom of the falls, and she warned me, "If you fall, I can't drag your ass back up." Thankfully, I didn't.

I found a third-party app that sends the drone to 400 feet and takes a series of 23 pictures. When I download the shots, I use another program to stitch them all together, creating an image that looks like it was taken from outer space. Our home in Vermont is now the Center of the Universe!

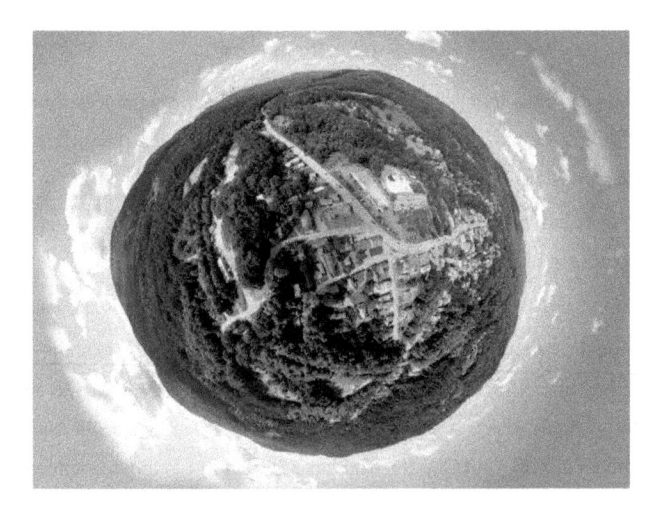

I was shooting some video over a covered bridge when a bald eagle flew under the drone. I grabbed a couple of dozen screenshots and created this composite image.

I'm told that birds of prey will attack a drone and that the best defensive maneuver is to power straight up. Fortunately, I didn't have to try that.

Indoor Skydiving

Penny has heard my skydiving stories for years but has never experienced freefall herself. We both signed up for a session of indoor skydiving in a wind tunnel near Baltimore. The skill of the experienced flyers is astonishing. Google "wind tunnel flyers."

When you're new to this, an instructor is always with you. He or she will even fly your way up to the top of the chamber.

Penny asked me what it was like to go up, and I replied, "I don't know. I've never fallen up."

I had hoped to do some flips and barrel rolls, but they required several hours of "flight" time before they'd allow that. They did let me fly free for quite a while, and it was fun to enjoy the experience after so many years. A week later, we were in Orlando and did it again. They will provide you with a video of your flight. If you haven't tried it, I highly recommend putting it on your bucket list.

Sailing

I mentioned earlier that some activities come easy to me and others do not. Sailing clearly falls in the latter column. My friend Gary is an avid sailor and encouraged me to try out his favorite pastime. He lives on a large nearby lake and has a few small sailboats. Google "sunfish" to get an idea of what they look like.

Gary gave me an introductory course and turned me loose. I managed to make my way across the lake in very light winds, but the bottom of the sail called the boom, kept swinging back and forth, hitting me in the head.

I headed back toward his dock when the winds on the appropriately named Lake Rescue suddenly picked up. My short lesson hadn't covered how to handle this situation, so instinctively, I pulled the rope tightening the sail. As I discovered, that was the exact *opposite* of what I needed to do.

I was wearing a GoPro in a chest harness and videoed the mayhem that ensued. The mast tilted to the left, and I pulled harder, causing it to drop even farther. It became abundantly evident that a capsize was inevitable, and sure enough, I found myself "swimming with the fishes."

I had been told that if the boat did capsize, I should put my weight on the centerboard to cause the boat to right itself. Unfortunately, the centerboard had come loose and was floating next to me! With a great deal of effort, I retrieved it and stuck it halfway through the slot at the bottom of the boat. I couldn't get enough leverage, but fortunately, Gary and his neighbor had witnessed my ineptitude and boated out to help me. The mast was stuck in the mud at the lake bottom, and it took a great deal of effort to "right the ship."

Sailing was added to the growing list of activities to avoid in the future…

Our Hot Air Balloon Misadventure

To celebrate our 20th anniversary, Penny announced that she wanted to take a hot air balloon ride. Nearby Quechee, Vermont, has an annual balloon festival that offers rides, so we signed up for it. Our pilot was highly experienced with 360 flights, but things didn't exactly go according to plan. Takeoff was borderline as the winds were higher than desired. It was a small basket with only Penny, the pilot, and myself. I was wearing a head-mounted GoPro to record our adventure. Right from the outset, the chase crew had trouble keeping up. We tried to land in a mobile home park, but there was no one to grab the tether rope, so it was back up again. When you fire flames into the balloon, it takes a long time to respond. As a result, we crashed the basket through trees seven times!

Daylight was disappearing, as were potential landing spots. In the distance, we could see the Connecticut River and New Hampshire beyond it. Airspace around the airport in Lebanon is a no-no for balloons, so that was a concern as well. I don't know how much fuel we had left, but the pilot was looking grim. We spotted a huge cornfield and, after one more crash through the trees, settled down into it.

The chase crew had no clue where we were. Finally, I had an idea. I did a screen capture of a Google map satellite view on my iPhone and texted it to them. Unfortunately, there were no visible roads leading into the cornfield. After an hour, they made their way into the far end of the field, about a quarter mile away. We had kept the balloon inflated because we needed to be towed to the vehicle. With one poor guy pushing and one pulling, we covered the distance wiping out stalks of corn. Penny joked that this was how crop circles were formed. Eventually, we made it and got the balloon deflated and stowed aboard the vehicle. Traditions can't be ignored, so we stopped for champagne on the way back to the launch point.

Would we do it again? You bet!

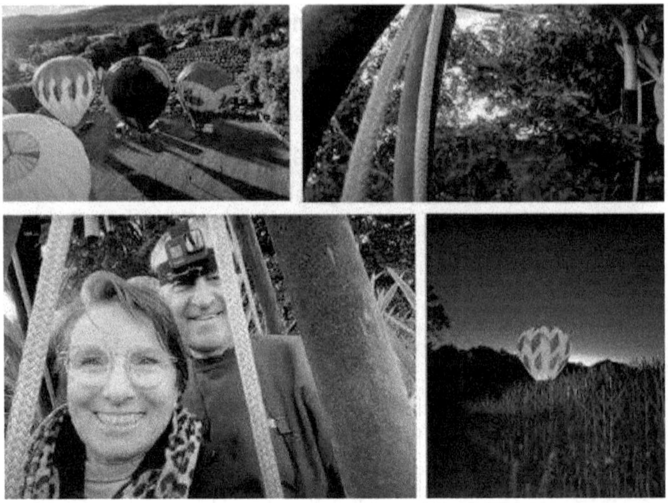

Retirement (sort of)

When I discover a new activity or pursuit, I'll sometimes jump in with both feet and devote a lot of time and energy. A decade ago, we visited an Albrecht Dürer exhibit in Germany, and I got the urge to take up pencil drawing. When we got back to the states, I learned the basics from a website and proceeded to crank out drawing after drawing. They got better over time, and I started to pick up some commissions. During the COVID-19 Lockdown, Penny signed me up for a three-day watercolor class on Zoom. I rapidly did about fifty of them. Penny then bought a starter kit for acrylic painting for herself, and of course, I couldn't resist trying them.

Over the past year, I completed more than 200 acrylic paintings. Ever since I retired, I can't sleep past 4 am. It's the perfect quiet time to paint. I gave away about fifty paintings and then discovered that folks would buy them. This past summer, I participated in a half dozen art and craft shows selling paintings, drawings, and photography. I've started shooting drone photography and discovered there's a market for that as well.

I designed the craft show display setup in the photo above and posted a "how to" video on Youtube. It's had nearly 50,000 views!

Life is good!

Epilogue

You've just learned about a few adventures and pursuits I've had over the past ¾ century. I have no intention of slowing down, and I'm now teaching myself to play the piano. We're booking more trips and are always looking for the next adventure.

At various times I've been a TV weatherman, gymnast, springboard diver, ski racer, snowboard racer, college instructor, USAF airborne weather observer, computer geek, insurance salesman, small business owner, law student, rock climber, golfer, recruiter, web designer, photographer, videographer, artist, flyboarder, bungee jumper, unicyclist, barefoot waterskier, parachute rigger, pilot, skydiver, and author. My motto: "If you're not living on the edge, you're taking up too much room."

Penny suggested that I put together a short video encapsulating some of the adventures described in this book. I used Bryan Adam's song "18 'Til I Die" for a backing track. Check it out at **bit.ly/pat1946**

Life in Vermont is good. I was asked recently if I've lived here my whole life. I replied, "Not yet!"

Our friend Phil Ayoub is a gifted singer/songwriter. One of the songs from his album Arrivals and Departures, "Get Out (Live a Little, Love a Lot)," has the following lyrics:

Get out

Get out the house

Out in the street

Get out in your town

Get off

Get off the couch

Get off line

And get on life

Bust out

And bet like you're leaving

Life ain't about

Just breaking even

Get out the house

Out in the street

Get out in your town

Get off

Get off the couch

Get off line

And get on life

Bust out

Cause life ain't about just breaking even

Live a little

Love a lot

Take a little chance it's all that you got

Live a little

And love a lot

Come on get out

Great advice. Get off the couch and make the most of *your* lives!

When you're over the hill, you pick up speed!

Not quite the end

I've had a lot of fun jotting my memories down in this book. By any measure, I'm lucky to be alive. What led to my proclivity for getting into dicey situations? I thought I'd turn back the clock and reminisce on some events in my formative years that might provide clues to my future behavior....

Let's go back a little bit in time. At noon on October 25th, 1946, I entered the world at Elkhorn Hospital in southern Wisconsin. Mom wanted her firstborn child to be delivered near her hometown of Delavan. Dad was a young Air Force Captain stationed at Tinker Air Force Base in Oklahoma, so five days later, I became an Okie. That lasted a year before we boarded a plane bound for Nagoya, Japan, with a stop in Anchorage. It was only two years after we had laid waste to Hiroshima and Nagasaki. Two years later, we set sail back for the states on the US Attack Transport ship James O'Hara. My earliest memories are from that voyage. On high seas, my high chair rocked backward, landing me on the deck. Ironically I repeated that move in a game of Dead Ants in a bar twenty years later.

We crossed the International Dateline on Christmas Day, setting the calendar back. Santa gave me books of Lifesavers on both days. Maybe the Lifesavers would be a sign that I'd be kept safe in years to come. We docked at Seattle on New Year's Eve 1949.

Never leave a six-year-old in the car unattended

When we got back to the states from Japan, Dad was stationed at Andrews AFB, and we bought a small cape-style house in the town of Forestville, MD. My memories of that time are a little clear. I recall getting my tonsils taken out, falling off a fence and severely gashing my left wrist on a broken bottle, and my eighteen-month-old brother burning his hand on a steam iron and running down the street totally naked. Most of all, I remember Dad's Mercury. This was his pride and joy, and justifiably so. It was a 1949 convertible in deep blue, and it actually had a power roof and power windows. He paid $3000 for that car, and it was a LOT of money back then. The car had one of those cool spotlights that we played with at the drive-in movie before the show started. When the roof was up, there was a fabric well behind the rear seat that we kids called the "way back."

In 1952 Dad was transferred to the Pentagon to work for the Joint Chiefs of Staff, and we moved to Arlington Forest, Virginia. Our Mercury was easily the coolest car in the neighborhood, and we loved to go out for rides in it. One Saturday, Dad had to pick up some paperwork from the Pentagon, so he and I hopped in the Mercury and headed out. He hadn't planned to be inside for very long, so he left me in the car with the usual admonition of keeping the doors locked and not talking to strangers. I was a very inquisitive kid and couldn't sit still for long. I was fascinated with cars and began exploring the interior of the Mercury. Above the rearview mirror, I found what looked like a door handle between the two visors. A six-year-old can't be expected to not test out every unfamiliar knob and handle, and I was no exception. I turned the knob, and it didn't do anything, so I lost interest in that and started playing with the cigarette lighter. Fortunately, I didn't burn anything and got it back in its receptacle just as Dad returned to the car.

We took off and headed for the highway. There wasn't much traffic, so we picked up speed quickly. We both sensed some movement overhead, and the entire convertible top suddenly shot straight up and froze in a vertical position like a giant sail. The "air brake" brought our speed down rapidly. We came to a halt, and both stared upward. Slowly our heads turned around and saw that the metal braces used to raise the roof were mangled and immovable. Clearly, the roof wouldn't retract into the way back, nor would it lower to its normal position. I looked at Dad, and he looked at me. He didn't say a word. He drove very slowly (you can only drive slowly when you have a giant wind anchor), and we inched our way down the road with people gawking, pointing, and laughing. We pulled into a garage, and all the employees stopped what they were doing to come out and stare at what I had done. Dad called Mom to come to get us with our other car, and Dad never brought up the subject again. I knew then that he truly loved me.

From the fall of 1956 until the spring of 1959, Dad had a 2 ½ year tour of duty as a weather officer at Wiesbaden AFB in Germany. It was an exciting time, and I really enjoyed myself. Most of my classmates had no interest in learning anything of the culture or language, but I made friends with a young German boy named Helmut, who lived on a sugar beet farm just off base, and he taught me a little German.

* * *

In January of 1958, our Boy Scout scoutmaster had the bright idea of holding a winter campout in the Taunus Mountains near Wiesbaden, Germany. I guess he thought it would teach us scouting is more than casual hikes, summer campfires, and Kumbayah. We went to Special Services at the Air Base and checked out shelter halves. Our less informed civilian friends called them pup tents. We headed into the mountains despite predictions of a snowfall of epic proportions. We were familiar with the campsite, having camped there the previous summer. Our campfire was built in the accustomed spot, and we sang the same silly songs we had sung six months earlier. When it was time to go to bed, we followed the time-honored Boy Scout tradition of peeing on the fire. It could have been extinguished with snow, but traditions must be upheld. Ladies, you have *no* idea how bad that smells.

To this day, I don't recall the name of my tent mate, but I suspect he remembers me. Please understand that I was very little for my age. As late as the ninth grade, I was the smallest kid, boy or girl, in my two-town high school. Back in 1957, when I was eleven, I was *really* small. My folks had bought me a sleeping bag the color of the Jolly Green Giant. It was bigger than it needed to be and was shaped like a mummy. It had a hood with a drawstring like a modern sweatshirt, and its insulation capabilities were marginal. Sometime during the night, the snow started to fall and accumulated two feet. The tent was only three feet tall, leaving a very small pyramid-shaped exit. The temperatures dropped precipitously, and I got cold. I pulled my head back into the sleeping bag, much as a turtle retracted his. Then, while asleep, apparently, I managed to do a 180 with the result that my feet were now pointed toward the hood end, and my head was where my feet were supposed to be.

At some point during the night, my bladder told me it was time to step outside. I recall groggily reaching out with my hands for the opening and discovering it wasn't there! I didn't panic at first but slowly ran my hand along the sleeping bag's seam in an attempt to find a way out. The only thought I had was that someone had come along in the middle of the night and had sewn up the opening to the sleeping bag. At that point, panic did set in. I had a hard time breathing and started kicking and clawing. When that did no good, I began rolling. I rolled over my tentmate, knocking down one of the support poles bringing down the tent with a lot of accumulated snow. Suddenly my foot stuck out the open hood and into wet snow. At the same time, my hand grazed the zipper, and I slowly undid it and slid out of the sleeping bag tail first. Hyperventilating, I finally emerged like a newborn in a breach delivery. My tentmate was staring wild-eyed at the apparition before him. I looked at the mayhem I had caused, decided it wasn't worth getting up, wet my sleeping bag, and went back to sleep.

We were introduced to skiing and took a trip to Bavaria with Major Cram and his family. The Cram kids were older, but their youngest son Steve was in my grade. This was the same Steve who had accompanied me on my assault on an abandoned cruise ship in Maryland ten years later. Small world. We were classmates in junior high in Germany and our senior year in Suitland, MD.

Dad was an avid skier and competitive ski jumper in Wisconsin in the 1930s, and he was anxious to introduce us to the sport. We spent Christmas 1956 at the Skytop Inn, which had been used as Hitler's stables during the war. Our first foray on the slopes was at a place called Rossfeld on the Austrian border. We used leather boots, wooden skis, and bamboo poles. We three kids took to the sport quickly. A decade later, my own military service interrupted my skiing, but my brother stayed with it. Even today, he's one of the best bump skiers I've ever seen. Back in the fifties, our technique was primitive but more than made up for by our enthusiasm.

Military brats have their own jargon. When you met someone new your age, you invariably asked, "When do you rotate?" Rotation was the term for ending a duty assignment and moving on to the next one. I'm frequently asked if it was tough always leaving friends, and the answer is no. We looked forward to the next assignment like it was a new adventure. Just in my twelve years of school, I lived in Virginia, Alabama, New York, Germany, Eastern Massachusetts, Alabama again, Western Massachusetts, and Maryland. I attended four different high schools.

Dad's mellowness is tested again

In any event, back in the fifties, we eagerly awaited news of our next duty assignment. In this case, Dad was stationed at Hanscom AFB near Bedford, MA. We bought a house in Sudbury, but it was under construction, and we needed a place to stay for about three months. Our realtor had arranged for us to rent a cottage on Scraggy Neck in Buzzards Bay out on Cape Cod. Dad was going to stay at the Bachelor Officer Quarters and join us on weekends. Mom would have our 1955 Chevy wagon while Dad would drive his 1956 Triumph TR3 sports car. One day in May of 1959, we all rode in the Chevy to the Cape to check out the cottage. The plan was to meet the realtor at a rotary not far from the causeway at a predetermined time.

We made a pretty good time driving down from Sudbury and stopped to eat at Howard Johnson's. For some reason, Dad had entrusted my nine-year-old brother with the car keys, which he left on the back seat as he went around locking all the car doors. We emerged from the restaurant with full bellies and little time to spare to get to our appointment. When Dad asked Mike for the keys, he fished into his pocket and came up empty-handed. My sister Jan spotted them on the back seat, and we all pressed our noses to the glass to confirm that they indeed were out of reach. This was long before the days of cell phones, and we had no way of getting to our appointment or contacting the realtor. Mom noticed that the car next to ours was the same make and model and said, "Don, why don't I just go back inside and borrow this car owner's keys"?

Dad started to say something and thought better of it. Mom took off on her errand as he tried to come up with a solution. Time was running off when he spotted a brick lying just off the parking lot. In desperation, he hefted it and took aim at that little vent window that cars used to have. He took a deep breath and smashed the brick into the glass. At that precise moment, Mom opened the passenger door with the borrowed key. As incredibly remote as the odds were, the key *was* the exact match as ours. We all just sort of stood there for what seemed like an interminable time. Mom finally excused herself to return the key. Dad cleaned up the broken glass from his seat. When Mom returned, we all got in the car, and nobody said a word. Dad slowly backed the car out of its parking place and headed for the highway. At the stop sign, what remained of the window fell out on the pavement. Mike, Jan, and I wanted to laugh, but somehow we sensed that wouldn't go over too well. The back of Dad's neck took on a kind of crimson hue we had never seen before. We rode in silence to our appointment and went on to rent the cottage. Dad returned to Hanscom

AFB, and Mom got the window fixed. We kids settled into a life of beach living for the next three months and finally moved into our new house in Sudbury.

In my entire life, I never saw my father lose his temper. There certainly were many times when he would have been justified, but he showed me that it's possible to take anything in stride.

Sleep disorders

I've suffered from Restless Leg Syndrome since the age of 19, and several members of my family are similarly affected, but the sleep "disorders" I'm about to discuss have nothing to do with neurological problems.

In the summer of 1965, I was home from my freshman year at the University of Florida on summer break. I was on a diving scholarship at the school that Playboy Magazine had dubbed the number one party school in America. In the preceding months, I had rushed a fraternity and developed a fondness for beer. We lived in Suitland, Maryland, at the time, where the drinking age was 21. In the nearby District of Columbia, beer could be consumed at the age of 18. One night my friends Alan, Steve, and I headed for Georgetown for a bar called The Cellar Door, where we consumed *way* too many beers. Long after we should have left, we stumbled back to Steve's car (I was the designated drinker) and headed home. Steve dropped me off at my Dad's house, and I managed to find my way downstairs to the basement bedroom I shared with my younger brother, Mike.

Since Mike was a full-time resident, he laid claim to the lower bunk. I managed to crawl into the upper and collapsed. Apparently, I remained essentially inert for a few hours. When I reconstructed the events much later, I deduced that my left arm had been pinned under the weight of my body the entire time and had absolutely no feeling. When I awoke in the middle of the night, I recall sensing that I was not alone in bed. Something warm and hairy had crawled into my bed and was directly under me. I was too groggy to panic and slowly reached my right hand under me to find out what I was sharing the bed with. My hand made contact with the lifeless arm, and that's when all rational thought and action went out the window. I grasped the offending limb and threw it as hard as I could at the wall. It bounced back off the wall and hit me in the chest.

I shrieked and dove backward off the edge of the upper bunk and hit the floor, landing on the middle of my back. Unfortunately, my feet were entangled in the sheets, which, in turn, were tightly wrapped around the mattress. Everything came down on top of me. Of paramount concern to me was that the "thing" had also landed on top of me. I began beating it fiercely with my right hand when a light was turned on. My bleary-eyed brother was staring at me, lying on the floor, pounding the bejeesus out of my left arm. All motion ground to a halt. I picked up the senseless limb, examined it, and dropped it. It was already turning black and blue. Without a word, Mike turned the light back off. I curled up in whatever bed covers had landed on me and went back to sleep. When morning came, we never discussed the incident. More than fifty years later, he's managed to suppress the memory, but it will live with me forever.

Bond, James Bond....

I had a job as a lifeguard at a big hotel in southeast DC one summer, and my buddy Steve would come downtown when my shift ended. We'd pick up a bottle of Manishevitz Kosher wine and head for Haines Point, and plan our next mischief. One day Steve seemed unable to contain himself, and when I asked him what he was grinning about, he produced a bottle of peroxide. Surfer dudes were very big that summer, and I figured my chances of getting dates with any of the guests at the hotel would increase exponentially if I were tanned *and* blond. We applied a little of the peroxide to our hair and checked the results in the car mirror. Nothing! So we added some more. With no discernable results, we headed for the Jefferson Memorial with our bottle of wine and the peroxide and sat down, basking in the bright sun of that beautiful day. We reasoned we had a weak concentration of peroxide and dumped the rest of the bottle on our heads before finishing off the wine. We headed to our respective homes, and when I walked in the door, my Dad immediately said, "What happened to your hair"?

I looked in the mirror and was stunned to see one of the Beach Boys staring back at me. Apparently, the peroxide just took a little while to take effect. I stammered, "There was a chlorine leak at the pool, and I uh….." I couldn't think of anything else to say, and Dad just went back to reading his book. A week later, my younger brother came home with hair as blond as mine. The female guests at the hotel *did* seem to show more interest. Life was good.

During that same summer, we were all abuzz about the James Bond books and movies. This was the coolest dude on the planet, and we all wanted to be secret agents. Steve, Alan, and I were not ones to sit back and dream. We acted! South of Washington, DC, the Wilson Steamship line had abandoned a ship in a wide estuary near the beltway. It was mired in shallow water a few hundred yards from shore. Our James Bond-like adventure would be to get out to the liner, explore it, retrieve a souvenir, and get home without being detected. With great planning, we accumulated wet suits, an inflatable air mattress, and waterproof flashlights. We set out in my Dad's 1961 Ford Falcon station wagon and headed for the town of Oxon Hill. We stopped in a residential neighborhood a few blocks from the water's edge and gathered up our gear. Being careful not to alert any of the residents, we quietly walked to the shore and blew up our air mattress. We climbed into the wetsuits and applied smudges to our faces. A few tools went into a waterproof bag. The three of us slid our upper bodies laterally across the air mattress and used it as a support as we slowly kicked off the shore. It was a warm night, and there wasn't much wind.

The trip out was uneventful, and we soon found ourselves alongside the hull. The ship was sitting so low in the water we had no difficulty getting aboard. Once on board, we began exploring and found that the ship had been picked over pretty well. I can't recall if we were even able to retrieve a souvenir. We headed back and were nearing shore when we spotted a police car parked about a block away from my car. We immediately changed course and emerged several hundred yards from the point we entered the water. We huddled and tried to figure out what James Bond would do. Since it was my Dad's car, I was elected to retrieve it and meet the others a few blocks away.

I stripped out of my wet suit and set out, keys in hand. I could see the cops standing by the shore with powerful flashlights. Fortunately, I had left the driver's window down. I didn't want the interior light coming on and alerting anyone of my presence. I was gambling that they hadn't figured out which car belonged to the trespassers. I climbed into the driver's seat and pushed the clutch pedal. I slipped the gear shift into 1st gear and turned the key. As soon as the engine caught, I popped the clutch and put all 85 horsepower to work getting out of there. A few blocks away, I hit the brakes, and the others climbed in. We beat feet out of Oxon Hill and got back to Hillcrest Heights, where Steve and Alan lived as quickly as we could. I changed back into dry clothes and headed for home. I wandered in and found my Dad watching TV. "What have you been up to"? he asked.

"Oh, nothing. We were just goofing off tonight."

"By any chance, were you anywhere near Oxon Hill"?

"Um, maybe…." "How did, um, why"?

"I got a call from the Oxon Hill police. One of the residents spotted flashlights on the old Wilson liner and called the cops. The only unidentified car in the neighborhood had a license plate that matched a car I own". Any idea how that could be"? he added. I've never been able to pull anything over on Dad, so I fessed up and told him what we had done and why.

He thought for a moment and then said, "James Bond, huh?" Hmmm…. He didn't say anything else, and he never mentioned the incident again.

Dad was a huge influence on my life, and his story is worth sharing. Growing up on a farm in Southern Wisconsin, his father died when Dad was only 12 years old. He dropped out of school for a year to help run the farm, but it was lost in the Great Depression. He enlisted in the Army Air Corps at the beginning of WWII and was singled out in his large class at Basic Training to attend Officer Candidate School. He was assigned to the Joint Chiefs of Staff at the Pentagon in the early fifties and sold real estate in the evenings.

Tragedy struck as our Mom died in an auto accident when I was fourteen. Dad was faced with raising four kids as young as nine months old. He stayed in the Air Force and remarkably (without a college degree or pilot's wings) made full Colonel before the age of 40. Growing up in the Depression taught him solid values and a healthy dose of common sense. Other than homes, he never financed anything in his life.

Upon retirement, he became one of the top officials at NOAA and testified before Congress on multiple occasions. After a second retirement, he became a very successful realtor, eventually settling in Stowe, VT. He and his second wife, Mary, traveled the world more than anyone else I know. They ultimately settled in Charleston, SC. Ironically, Dad took up flying after retirement from the Air Force, eventually owning a couple of Cessnas. I'll always be grateful that he introduced me to alpine skiing, water skiing, skydiving, and golf and inspired us to travel as much as possible.

Mary died in 2018 at age 97. A year later, Dad passed away just a few weeks short of his 98th birthday. He was buried with full military honors at Arlington National Cemetery. Dad was the best role model we could ever have hoped for. The world lost an extraordinary man.

THE END

Lightning Source UK Ltd.
Milton Keynes UK
UKHW020439211122
412554UK00016B/836